Oklahoma Mammalogy

AN OKLAHOMA MUSEUM OF NATURAL HISTORY PUBLICATION

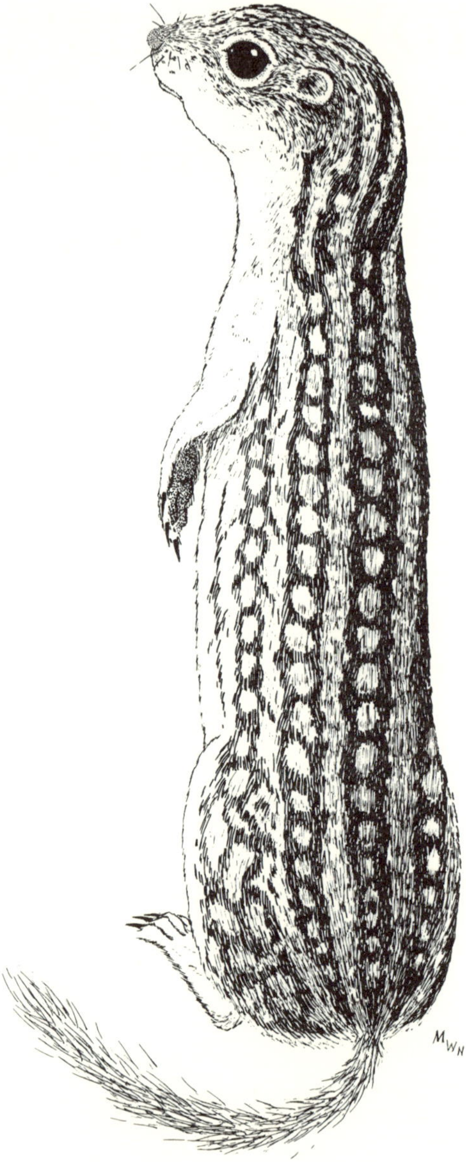
MWN

OKLAHOMA MAMMALOGY

AN ANNOTATED BIBLIOGRAPHY AND CHECKLIST

Robert D. Owen
and Gary D. Schnell

University of Oklahoma Press · Norman and London

The cover and frontispiece drawing of a thirteen-lined ground squirrel (*Spermophilus tridecemlineatus*) is by Michael W. Nickell.

Library of Congress Cataloging-in-Publication Data

Owen, Robert D.
Oklahoma mammalogy : an annotated bibliography and checklist / Robert D. Owen, and Gary D. Schnell.—1st ed.
p. cm.
"An Oklahoma Museum of Natural History publication."
Includes index.
ISBN 0-8061-2185-8 (alk. paper)
1. Mammals—Oklahoma—Bibliography. I. Schnell, Gary D. (Gary Dean), 1942- . II. Title.
Z7996.M3094 1989
[QL719.05]
016.59909766—dc19 88-27959

First edition.

The paper in this book meets the guidelines for permanence and durability of the Committee on Production Guidelines for Book Longevity of the Council on Library Resources, Inc.

To Susan, Bill, Maria, and Rabbit
and
Mary Sue, Steve, and Phil

For all the night hours
we missed at home
while preparing this work

Contents

Preface

A work of this type requires the assistance of many individuals. The late J. Keever Greer, who was the Director of the Stovall Museum of Science and History, provided initial encouragement and assistance with the project. Numerous helpful suggestions were offered at all stages by Joseph A. Grzybowski. Janet K. Braun, William Caire, Richard L. Cifelli, Michael E. Douglas, Charles M. Drabek, Michael A. Mares, Byron B. Moser, Harold E. Namminga, Greg L. Summers, Jack D. Tyler, and Don G. Wycoff furnished information on many of the references included in this work. Paul A. Vohs, Jr., Frank Schitoskey, and O. Eugene Maughan, Unit Leaders of the Oklahoma Cooperative Fish & Wildlife Unit, provided access to the Unit's files and publications library.

Sandra J. Herrington, Marcy L. McConnell, and Lisa Y. Pierce assisted with the verification and annotation of a number of references. Useful suggestions concerning the manuscript were provided by J. Knox Jones, Jr., and an anonymous reviewer. Michael W. Nickell provided the drawing for the cover and frontispiece.

Patricia Mills of the Bizzell Library's Interlibrary Loan Department was extremely helpful in obtaining publications for us. A number of individuals were involved in data entry and editing, including Cheryl D. Davis, Varga L. Knight, Melinda S. West, Lisa J. Cox, Sonya E. Johnson, J. Curtis Creighton, Valerie Y. Guliford, and Diane L. Fields. Much of the computer software used in processing the bibliography file was created by Daniel J. Hough and Thomas M. McKenna. Dan was largely responsible for developing and managing the computer files for the project. He also designed a number of computer procedures to assist us in the detecting of errors and inconsistencies. Diane L. Fields, Richard L. Cifelli, James J. Krupa, Michael A. Mares, Gary W. Shugart, and Janet L. Leonard helped with proofreading. John N. Drayton of the University of Oklahoma Press showed patience and provided encouragement throughout the project. To all of these persons, we extend our gratitude.

Financial assistance for this project was provided by the Oklahoma Biological Survey, the Oklahoma Museum of Natural History (through Director Michael A. Mares), and the Office of Research Administration (through Vice Provost Kenneth L. Hoving) of the University of Oklahoma.

Norman, Oklahoma

ROBERT D. OWEN

GARY D. SCHNELL

Oklahoma Mammalogy

Introduction

As a result of the central geographic location of Oklahoma, as well as its varied topography and large size, the mammal life of the state is rich and diverse (Blair and Hubbell, 1938; Hellack, 1974). The fauna and flora of Oklahoma include what typically are considered eastern and western components, as well as both northern and southern species (Blair, 1939; Duck and Fletcher, 1944); because of the midcontinental position of the state, a number of taxa come into contact that do not meet over most of the rest of their respective ranges. Thus, mammalogists and other biologists have found the region to be of special research interest.

Oklahoma became a state in 1907. From years prior to that date, we have the published observations of members of several surveying and hunting expeditions that traversed the region. The accounts of these early authors—for example, S. F. Baird, D. B. R. Keim, R. B. Marcy, C. H. Merriam, and S. W. Woodhouse—indicate that game and furbearing mammals were abundant. The first checklist of Oklahoma mammals was published in 1917 by Cross. Later, Blair and Hubbell (1938) and Blair (1939) produced lists of Oklahoma mammals, describing their distributions, habitats, and faunal associations. Hellack (1974) provided an updated checklist, which also indicated approximate distributions.

Although bibliographies of mammalogy exist for Texas (Raun, 1962; Jones and Homan, 1976; Jones et al., 1985) and some other nearby states, no such compilation has been published previously for Oklahoma. Our listing, which includes 1,208 citations, reflects the fact that, since the turn of the century, the number of publications concerning Oklahoma mammals has grown dramatically, with the total doubling approximately every 15 years.

We have compiled this annotated bibliography of Oklahoma mammalogy to provide easy reference for persons interested in conducting research and surveys on Oklahoma mammals, and for others needing access to information on the fauna of the state. The bibliography contains all of the references to Oklahoma mammals through 1986 that we judged to be germane. We systematically searched several regional journals (*Proceedings of the Oklahoma*

Academy of Science, Southwestern Naturalist, and *Bulletin of the Oklahoma Ornithological Society*), as well as the *Journal of Mammalogy, North American Fauna, Mammalian Species,* and the *Journal of Wildlife Management.* We also made use of a number of bibliographic sources, including *Zoological Record, Wildlife Review, The Mammals of North America* (Hall and Kelson, 1959; Hall, 1981), *Oklahoma Archaeology: An Annotated Bibliography* (Bell, 1969, 1978), and the Recent Literature supplements to the *Journal of Mammalogy.*

In addition, we have included dissertations and theses reporting on research conducted in Oklahoma, Pittman-Robertson Reports of the Oklahoma Department of Wildlife Conservation, and reports of the International Biological Program, as well as numerous miscellaneous books, monographs, and other publications. The bibliography includes a sampling of the popular literature; typically, we included only those articles that were of potential scientific or historical interest.

A number of the listed reports of the Department of Wildlife Conservation are not formal publications. However, we have included them here because they often contain data and conclusions that will be of considerable value to those interested in Oklahoma mammals.

The completeness of the bibliography is difficult to determine. References containing only casual mention of Oklahoma mammals generally were not listed. However, if we were in doubt concerning the potential significance of a reference to Oklahoma mammalogy, it was included. For the major mammalogical, regional, and wildlife journals, as well as dissertations and theses from Oklahoma State University and the University of Oklahoma, we believe the bibliography is essentially complete.

We found many pertinent reference publications by scanning literature citations in previously located publications and by obtaining the curricula vitae of authors known to have worked in Oklahoma. We corresponded with a number of individuals concerning publications on Oklahoma mammals and also located some works purely by chance. Undoubtedly, some dissertations and theses from non-Oklahoma institutions, as well as miscellaneous contributions in a variety of types of publications, were overlooked. Nonetheless, the references included here form the primary published basis of our knowledge of Oklahoma mammals.

Each reference in the main listing of the bibliography includes the author, date of publication, title, journal or book (editor, if publication is part of a book or monograph containing more than a single work), and publisher (unless published in journal or other series). References do not have the issue number indicated for consecutively paged periodical volumes, nor have we included Roman-numbered pages. We have routinely italicized scientific names for genera and species in titles. An annotation that describes the contents of the reference also is included. Each reference has been assigned a number, based on the sequential alphabetical listing, for use in the index that follows the main body of the bibliography.

We selected the following series of standardized terms to be included, where appropriate, in the annotations of a given reference as well as in the Family and Key Word Index: systematics, paleontology (including archaeology), history, distribution, natural history, behavior, ecology, populations, parasites, genetics, morphology, physiology, techniques, economics, wildlife management, popular, and bibliographies. Each annotation contains one or more key words.

In addition, each listing includes at least one mammal family name, unless a given reference has to do with more than six families of mammals; for these publications, we have used the designation "Mammalia" in the annotation. In a few cases, we included more than six families and, where a reference dealt primarily with a specific group or groups of mammals but more than six families were mentioned, the annotation includes the names of the primary families along with the designation "Mammalia." Family, genus, and species names (see Table 1 starting on page 8) follow the nomenclature of Jones et al. (1986) for extant species and Anderson and Jones (1984) for families represented only by fossil material.

The mammal families are listed alphabetically in the Family and Key Word Index, with appropriate key words and associated reference numbers being included for each family. The English Name Index at the end of the book indicates page numbers in the Family and Key Word Index where references for a particular mammal are listed.

In the bibliographic listings, abbreviations of journal names follow the American National Standard and are summarized in *Serial Sources for the BIOSIS Data Base* (BioScience Information Service, 2100 Arch Street, Philadelphia, Pennsylvania). Citations to Federal Aid Projects conducted by the Oklahoma Department of Wildlife Conservation indicate the job numbers and project code (and sometimes study number) along with the number of pages. We obtained listings of relevant reports through the Oklahoma City office of the Oklahoma Department of Wildlife Conservation (1979-86) and from the Fish and Wildlife Reference Service (pre-1979; P. O. Box 6044, Rockville, Maryland 20850). Photocopies of selected reports can be acquired from the Reference Service. The reports produced since 1979 are on file in Oklahoma City at the Oklahoma Department of Wildlife Conservation.

We have included citations to a number of publications on Oklahoma fossil mammals. Readers interested in the literature concerning fossil mammals in Oklahoma also are directed to the "Bibliography of Fossil Vertebrates" series authored by C. L. Camp, J. T. Gregory, and others. Papers in this series were published as: *Geological Society of America Special Papers* numbered 27 (covering 1928-33) and 42 (1934-38); *Geological Society of America Memoirs* numbered 37 (1939-43), 57 (1944-48), 84 (1949-53), 92 (1954-58),

117 (1959-63), 134 (1964-68), and 141 (1969-72); special volumes by the American Geological Institute (for 1973-77, 1978, 1979, and 1980); and annual volumes by the Society of Vertebrate Paleontology (covering 1981-84). Also, although a substantial sampling is included from those works in the Oklahoma archaeology literature where mammals are reported, we undoubtedly have not included an exhaustive list of such reports.

As one might expect for a regionally oriented bibliography, about half of the 1,208 references (603) are concerned, at least in part, with mammal distributions. References dealing with some aspect of ecology or natural history of Oklahoma mammals are well represented with 490 and 484 references, respectively. Six of the other topics were covered in at least 200 references: behavior, 295; morphology, 358; paleontology, 243; populations, 271; systematics, 314; and wildlife management, 243.

Among the families of mammals, those containing huntable wildlife species, predators, or agricultural pest species are most often the topic of publications as indicated below:

Family	English Name	Number of References
Cricetidae	Cricetids	216
Cervidae	Cervids	168
Sciuridae	Squirrels	134
Canidae	Canids	125
Vespertilionidae	Vespertilionid bats	102
Leporidae	Hares and Rabbits	90
Bovidae	Bovids	82
Mustelidae	Mustelids	78
Felidae	Cats	73
Heteromyidae	Heteromyids	72

The family Cricetidae includes a number of species considered to be agriculturally destructive. However, the relatively large number of references also reflects the large number of cricetid species (24) found in Oklahoma. Another frequently referenced, well represented family is Vespertilionidae, with 20 of the 108 mammal species in the state (Table 1) being bats of this family.

The periodical with the largest number of papers concerning Oklahoma mammalogy is the *Journal of Mammalogy* (150). Many references also were found in the *Proceedings of the Oklahoma Academy of Science* (92) and the *Southwestern Naturalist* (56). There were 87 relevant articles printed in *Outdoor Oklahoma* and 85 project reports of the Oklahoma Department of Wildlife Conservation. A total of 38 Ph.D. dissertations and 60 master's theses are included in the bibliography. Over half (58) of the Recent mammal species of Oklahoma have been covered to date in the *Mammalian Species* series published by the American Society of Mammalogists, and these publications are included in the references.

Obviously, numerous additions to the mammalogical literature of Oklahoma will be made in the future. To accommodate updating and revising this bibliography, we have stored and processed it at the University of Oklahoma using GIPSY (General Information Processing System), a general-purpose computer data storage-and-retrieval system. Thus, the bibliography can easily be sorted using key words, mammal family names, or other information contained in the included citations.

For typesetting, the bibliography was transferred to a personal computer. We employed WordStar 2000+™ software and a Hewlett-Packard LaserJet Series II™ printer to produce camera-ready copy.

Table 1. *Orders, families, and extant (or recently extirpated) species of mammals found in Oklahoma. Sequence, scientific nomenclature, and English names follow Jones et al. (1986) for extant species, with Anderson and Jones (1984) being used for extinct taxa.*

ORDER MARSUPIALIA (Marsupials)

DIDELPHIDAE (Opossums)

Didelphis virginiana (Virginia Opossum)

ORDER INSECTIVORA (Insectivores)

SORICIDAE (Shrews)

Blarina carolinensis (Southern Short-tailed Shrew)
Blarina hylophaga (Elliot's Short-tailed Shrew)
Cryptotis parva (Least Shrew)
Notiosorex crawfordi (Desert Shrew)

TALPIDAE (Moles)

Scalopus aquaticus (Eastern Mole)

ORDER CHIROPTERA (Bats)

VESPERTILIONIDAE (Vespertilionid Bats)

Myotis austroriparius (Southeastern Myotis)
Myotis ciliolabrum (Western Small-footed Myotis)
Myotis grisescens (Gray Myotis)
Myotis keenii (Keen's Myotis)
Myotis leibii (Eastern Small-footed Myotis)
Myotis lucifugus (Little Brown Myotis)
Myotis sodalis (Indiana or Social Myotis)
Myotis velifer (Cave Myotis)
Myotis yumanensis (Yuma Myotis)
Lasiurus borealis (Red Bat)
Lasiurus cinereus (Hoary Bat)
Lasiurus seminolus (Seminole Bat)
Lasionycteris noctivagans (Silver-haired Bat)
Pipistrellus hesperus (Western Pipistrelle)
Pipistrellus subflavus (Eastern Pipistrelle)
Eptesicus fuscus (Big Brown Bat)
Nycticeius humeralis (Evening Bat)

Table 1. [*continued*]

VESPERTILIONIDAE (Vespertilionid Bats) [*continued*]

Plecotus rafinesquii (Rafinesque's Big-eared Bat)
Plecotus townsendii (Townsend's Big-eared Bat)
Antrozous pallidus (Pallid Bat)

MOLOSSIDAE (Molossid Bats)

Tadarida brasiliensis (Brazilian Free-tailed Bat)
Tadarida macrotis (Big Free-tailed Bat)

ORDER XENARTHRA (Xenarthrans)

DASYPODIDAE (Armadillos)

Dasypus novemcinctus (Nine-banded Armadillo)

†GLYPTODONTIDAE (Glyptodonts)

†MYLODONTIDAE (Mylodonts)

†MEGALONYCHIDAE (Ground Sloths)

ORDER LAGOMORPHA (Lagomorphs)

LEPORIDAE (Hares and Rabbits)

Sylvilagus aquaticus (Swamp Rabbit)
Sylvilagus audubonii (Desert Cottontail)
Sylvilagus floridanus (Eastern Cottontail)
Lepus californicus (Black-tailed Jackrabbit)

ORDER RODENTIA (Rodents)

†MYLAGAULIDAE (Mylagaulids)

SCIURIDAE (Squirrels)

Tamias quadrivittatus (Colorado Chipmunk)
Tamias striatus (Eastern Chipmunk)
Marmota monax (Woodchuck)
Spermophilus spilosoma (Spotted Ground Squirrel)
Spermophilus tridecemlineatus (Thirteen-lined Ground Squirrel)
Spermophilus variegatus (Rock Squirrel)
Cynomys ludovicianus (Black-tailed Prairie Dog)

Table 1. [*continued*]

SCIURIDAE (Squirrels) [*continued*]

Sciurus carolinensis (Gray Squirrel)
Sciurus niger (Fox Squirrel)
Glaucomys volans (Southern Flying Squirrel)

GEOMYIDAE (Pocket Gophers)

Geomys breviceps (Baird's Pocket Gopher)
Geomys bursarius (Plains Pocket Gopher)
Cratogeomys castanops (Yellow-faced Pocket Gopher)

HETEROMYIDAE (Heteromyids)

Perognathus flavescens (Plains Pocket Mouse)
Perognathus flavus (Silky Pocket Mouse)
Chaetodipus hispidus (Hispid Pocket Mouse)
Dipodomys elator (Texas Kangaroo Rat)
Dipodomys ordii (Ord's Kangaroo Rat)

CASTORIDAE (Beavers)

Castor canadensis (Beaver)

CRICETIDAE (Cricetids)

Oryzomys palustris (Marsh Rice Rat)
Reithrodontomys fulvescens (Fulvous Harvest Mouse)
Reithrodontomys humulis (Eastern Harvest Mouse)
Reithrodontomys megalotis (Western Harvest Mouse)
Reithrodontomys montanus (Plains Harvest Mouse)
Peromyscus attwateri (Texas Mouse)
Peromyscus boylii (Brush Mouse)
Peromyscus difficilis (Rock Mouse)
Peromyscus gossypinus (Cotton Mouse)
Peromyscus leucopus (White-footed Mouse)
Peromyscus maniculatus (Deer Mouse)
Peromyscus pectoralis (White-ankled Mouse)
Peromyscus truei (Piñon Mouse)
Ochrotomys nuttalli (Golden Mouse)
Baiomys taylori (Northern Pygmy Mouse)
Onychomys leucogaster (Northern Grasshopper Mouse)
Sigmodon hispidus (Hispid Cotton Rat)
Neotoma albigula (White-throated Woodrat)
Neotoma floridana (Eastern Woodrat)

Table 1. [*continued*]

CRICETIDAE (Cricetids)[*continued*]

Neotoma mexicana (Mexican Woodrat)
Neotoma micropus (Southern Plains Woodrat)
Microtus ochrogaster (Prairie Vole)
Microtus pinetorum (Woodland Vole)
Ondatra zibethicus (Muskrat)

MURIDAE (Murids)

**Rattus norvegicus* (Norway Rat)
**Rattus rattus* (Black Rat)
**Mus musculus* (House Mouse)

ZAPODIDAE (Jumping Mice)

Zapus hudsonius (Meadow Jumping Mouse)

ERETHIZONTIDAE (New World Porcupines)

Erethizon dorsatum (Porcupine)

MYOCASTORIDAE (Myocastorids)

**Myocastor coypus* (Nutria)

ORDER CARNIVORA (Carnivores)

CANIDAE (Canids)

Canis latrans (Coyote)
Canis lupus (Gray Wolf)
Canis rufus (Red Wolf)
Vulpes velox (Swift Fox)
Vulpes vulpes (Red Fox)
Urocyon cinereoargenteus (Gray Fox)

URSIDAE (Bears)

Ursus americanus (Black Bear)
Ursus arctos (Grizzly or Brown Bear)

PROCYONIDAE (Procyonids)

Bassariscus astutus (Ringtail)
Procyon lotor (Raccoon)

Table 1. [*continued*]

MUSTELIDAE (Mustelids)

Mustela frenata (Long-tailed Weasel)
Mustela nigripes (Black-footed Ferret)
Mustela vison (Mink)
Taxidea taxus (Badger)
Spilogale gracilis (Western Spotted Skunk)
Spilogale putorius (Eastern Spotted Skunk)
Mephitis mephitis (Striped Skunk)
Conepatus mesoleucus (Hog-nosed Skunk)
Lutra canadensis (River Otter)

FELIDAE (Cats)

Felis concolor (Mountain Lion)
Felis rufus (Bobcat)

†ORDER PROBOSCIDEA (Proboscideans)

†GOMPHOTHERIIDAE (Gomphotheriids)

†ELEPHANTIDAE (Elephants and Mammoths)

†§MASTODONTIDAE (Mastodons)

†ORDER PERISSODACTYLA (Odd-toed Ungulates)

†EQUIDAE (Horses)

†TAPIRIDAE (Tapirs)

†RHINOCEROTIDAE (Rhinoceroses)

ORDER ARTIODACTYLA (Even-toed Ungulates)

†TAYASSUIDAE (Peccaries)

†CAMELIDAE (Camels)

†DROMOMERYCIDAE (Dromomerycids)

CERVIDAE (Cervids)

Cervus elaphus (Wapiti or Elk)
Odocoileus hemionus (Mule or Black-tailed Deer)
Odocoileus virginianus (White-tailed Deer)

Table 1. [*continued*]

ANTILOCAPRIDAE (Pronghorn)

Antilocapra americana (Pronghorn)

BOVIDAE (Bovids)

Bison bison (Bison)

† An order or family represented only by fossil species in Oklahoma.

* An introduced species in North America.

§ Mastodontidae has priority over Mammutidae (which has been employed in classifications used by a number of authors).

Annotated Bibliography

Annotated Bibliography

Ables, E., *see* 946.

1 **Ahshapanek, D. C., and R. D. Burns.** 1959. Mammals associated with prehistoric people of Oklahoma. Proc. Okla. Acad. Sci. 40:16-19.—Table of osteological material. Garvin County. Speculation on usage as food. Paleontology. Mammalia.

2 **Al-Jiboori, N. A. J.** 1968. Description and distribution of the integumentary glands of the guano bat, *Tadarida brasiliensis mexicana* (Saussure). M.S. Thesis, Oklahoma State Univ., Stillwater. 39 pp.—Study of integumentary histology of *Tadarida brasiliensis* from Major County. Morphology, Techniques. Molossidae.

Allen, G. M., *see* 801.

3 **Allen, H.** 1864. Monograph of the bats of North America. Smithson. Misc. Collect. 165:1-85.—Systematics, Distribution, Natural history, Parasites, Morphology. Vespertilionidae, Molossidae.

4 **Allen, J. A.** 1877. History of the American bison, *Bison americanus.* Annu. Rep. U.S. Geol. Surv. for 1875 9:443-590.—Includes previous and contemporary distributions of *Bison americanus* (=*bison*) in Indian Territory. Distribution. Bovidae.

5 **Allen, J. A.** 1877. Sciuridae. Monogr. N. Am. Rodentia No. 11:631-949.—Includes references to fox squirrel (*Sciurus niger*) specimens from Oklahoma and Indian territories. Systematics, Distribution, Natural history. Sciuridae.

6 **Allen, J. A.** 1891. Notes on new or little-known North American mammals, based on recent additions to the collection of mammals in the American Museum of Natural History. Bull. Am. Mus. Nat. Hist. 3:263-310.—Type locality of *Dipodops richardsoni* (=*Dipodomys ordii richardsoni*) on Beaver River in Indian Territory. Also specimens of *Neotoma micropus canescens* from same locality. Systematics, Distribution. Heteromyidae, Cricetidae.

7 **Allen, J. A.** 1901. A preliminary study of the North American opossums of the genus *Didelphis.* Bull. Am. Mus. Nat. Hist. 14:149-188.—Revisionary study of *Didelphis* includes specimens of *D. virginiana* from Oklahoma and Indian territories. Systematics, Distribution. Didelphidae.

Allen, J. A., *see* 265.

8 **Allen, P. F., and B. Osborn.** 1954. Tall grass defeats prairie dogs. Soil Conserv. 20:103-105,113.—Population of *Cynomys ludovicianus* in Comanche County declined and vanished, due apparently to revegetation of tallgrass prairie climax. Behavior, Ecology, Popular. Sciuridae.

Allen, P. F., *see* 850.

Allison, J. E., *see* 697.

9 **Allred, B. W.** 1947. The longhorn of the Wichitas. Cattleman 34:120-123.—Discussion of history and morphology of longhorn cattle (*Bos taurus*) in Wichita Mountains Wildlife Refuge, Comanche County. History, Behavior, Morphology, Wildlife management. Bovidae.

10 **Anderson, A. D.** 1962. The Cooperton mammoth: A preliminary report. Plains Anthropol. 7:110-112.—Description of Kiowa County *Mammuthus* site. Paleontology. Elephantidae.

11 **Anderson, A. D.** 1962. The Domebo mammoth kill site in Caddo County, Oklahoma. Great Plains Newsl. 2(2):4-6.—Preliminary report on Caddo County *Mammuthus* specimen. Paleontology, Popular. Elephantidae.

12 **Anderson, A. D.** 1975. The Cooperton mammoth. An early man bone quarry. Great Plains J. 14:130-173.—Extensive discussion of *Mammuthus* from Kiowa County site. Paleontology, History. Elephantidae.

13 **Anderson, A. D., and F. L. Chappabitty.** 1968. A bison kill from Domebo Canyon. Great Plains J. 8:48-52.—*Bison bison* skeleton found in Caddo County. Paleontology. Bovidae.

14 **Anderson, A. E., and O. C. Wallmo.** 1984. *Odocoileus hemionus.* Mamm. Species 219:1-9.—Summary of published information on mule deer, *Odocoileus hemionus.* Systematics, Paleontology, History, Distribution, Natural history, Behavior, Ecology, Populations, Parasites, Genetics, Morphology, Economics, Bibliographies. Cervidae.

Anderson, E., *see* 694.

15 **Anderson, S., and R. Barlow.** 1978. Taxonomic status of *Cervus elaphus merriami* (Cervidae). Southwest. Nat. 23:63-70.—*Cervus merriami* a subspecies of *C. elaphus.* Unable to assess taxonomic status of Oklahoma populations. Systematics, Distribution. Cervidae.

16 **Anderson, S., J. K. Doutt, and J. S. Findley.** 1963. Collections of mammals in North America. J. Mammal. 44:471-500.—Includes collections in Oklahoma (University of Oklahoma Stovall Museum, Oklahoma State University, Southwestern State College, University of Oklahoma Biological Station, and Wichita Mountains Wildlife Refuge). Some information on mammal families and localities of specimens. Systematics, History, Distribution. Mammalia.

17 **Anderson, S., and J. K. Jones, Jr.** 1967. Recent mammals of the world—A synopsis of families. Ronald Press Co., New York. 453 pp. —Discussion of families and genera. Systematics, Paleontology, Distribution, Behavior, Ecology. Mammalia.

18 **Anderson, S., and J. K. Jones, Jr.** 1984. Orders and families of Recent mammals of the world. John Wiley and Sons, New York. 686 pp.—Updated version of Anderson and Jones (1967). Discussion of orders, families, and genera. Systematics, Paleontology, Distribution, Behavior, Ecology. Mammalia.

Anderson, S., *see* 474, 641, 741.

19 **Anonymous.** 1959. State areas receive deer. Okla. Wildl. 15(3):5.—*Odocoileus virginianus* from Wichita Mountains Wildlife Refuge, Comanche County, transplanted to other areas in Oklahoma. Wildlife management. Cervidae.

20 **Anonymous.** 1962. State to have season on elk. Okla. Wildl. 18(5):9. —Free-ranging *Cervus elaphus* in Kiowa and Comanche counties. Distribution, Wildlife management, Popular. Cervidae.

21 **Anonymous.** 1973. Threatened wildlife of the United States. U.S. Fish Wildl. Serv. Resour. Publ. 114:1-289.—Distributions of *Myotis sodalis* and *Plecotus townsendii ingens* include portions of Oklahoma. Several other species covered may also occur in state. History, Distribution, Populations, Wildlife management. Vespertilionidae, Mammalia.

22 **Anonymous.** 1974. Wildlife portrait: The elk. Outdoor Okla. 30(9): 20.—*Cervus elaphus*. Account for the layman. Distribution, Natural history, Behavior, Ecology, Morphology, Wildlife management. Cervidae.

23 **Anonymous.** 1974. Wildlife portrait: The porcupine. Outdoor Okla. 30(11):20.—*Erethizon dorsatum*. Account for the layman. Distribution, Natural history, Behavior, Ecology, Morphology, Wildlife management, Popular. Erethizontidae.

24 **Anonymous.** 1974. Wildlife portrait: The skunk. Outdoor Okla. 30(7):20.—*Mephitis mephitis* and *Spilogale putorius.* Account for the layman. Distribution, Natural history, Behavior, Ecology, Morphology, Popular. Mustelidae.

25 **Anonymous.** 1974. Wildlife portrait: Tree squirrels. Outdoor Okla. 30(8):24.—*Sciurus niger* and *S. carolinensis.* Account for the layman. Natural history, Behavior, Ecology, Morphology, Popular. Sciuridae.

26 **Anonymous.** 1975. Wildlife portrait: The buffalo. Outdoor Okla. 31(3):20.—*Bison bison.* Account for the layman. History, Distribution, Natural history, Ecology, Morphology, Wildlife management, Popular. Bovidae.

27 **Anonymous.** 1975. Wildlife portrait: The gray fox. Outdoor Okla. 31(2):20.—*Urocyon cinereoargenteus.* Account for the layman. Systematics, Distribution, Natural history, Behavior, Ecology, Morphology, Popular. Canidae.

28 **Anonymous.** 1975. Wildlife portrait: Ground squirrel. Outdoor Okla. 31(4):20.—*Spermophilus tridecemlineatus, S. spilosoma,* and *S. variegatus.* Account for the layman. Distribution, Natural history, Behavior, Ecology, Morphology, Popular. Sciuridae.

29 **Anonymous.** 1975. Wildlife portrait: The swamp rabbit. Outdoor Okla. 31(11):18.—*Sylvilagus aquaticus.* Account for the layman. Systematics, Distribution, Natural histor, Behavior, Ecology, Morphology, Popular. Leporidae.

30 **Anonymous.** 1976. Wildlife portrait: The flying squirrel. Outdoor Okla. 32(11):20.—*Glaucomys volans.* Account for the layman. Distribution, Natural history, Behavior, Ecology, Morphology, Popular. Sciuridae.

31 **Anonymous.** 1976. Wildlife portrait: The jackrabbit. Outdoor Okla. 32(10):20.—*Lepus californicus.* Account for the layman. Systematics, Distribution, Natural history, Behavior, Ecology, Morphology, Popular. Leporidae.

32 **Anonymous.** 1976. Wildlife portrait: The raccoon. Outdoor Okla. 32(6):20.—*Procyon lotor.* Account for the layman. Distribution, Natural history, Behavior, Ecology, Morphology, Popular. Procyonidae.

33 **Anonymous.** 1977. Endangered and threatened wildlife and plants. Proposed endangered listing and critical habitat determination for the Virginia and Ozark big-eared bats. Fed. Reg. 42(232):61290-

61292.—Distribution of endangered bat *Plecotus townsendii ingens* includes eastern Oklahoma. Distribution, Populations, Wildlife management. Vespertilionidae.

34 **Anonymous.** 1977. Wildlife portrait: The gopher. Outdoor Okla. 33(6):20.—*Geomys bursarius.* Account for the layman. Distribution, Natural history, Behavior, Ecology, Morphology, Economics, Popular. Geomyidae.

35 **Anonymous.** 1977. Wildlife portrait: The mule deer. Outdoor Okla. 33(2):20.—*Odocoileus hemionus.* Account for the layman. Distribution, Natural history, Behavior, Ecology, Morphology, Popular. Cervidae.

36 **Anonymous.** 1977. Wildlife portrait: The muskrat. Outdoor Okla. 33(11):20.—*Ondatra zibethicus.* Account for the layman. Natural history, Behavior, Ecology, Morphology, Popular. Cricetidae.

37 **Anonymous.** 1978. Bobcat, lynx, otter, ginseng '78-'79 exports proposed. Endangered Species Tech. Bull. 3(8):10-11.—Population and harvest figures by state (1977-78) for *Lynx* (=*Felis*) *rufus* and *Lutra canadensis.* Wildlife management. Mustelidae, Felidae.

38 **Anonymous.** 1978. Export quotas set for bobcat, river otter, lynx, ginseng. Endangered Species Tech. Bull. 3(4):7-8.—Population and harvest figures by state (1976-77) for *Lynx* (=*Felis*) *rufus* and *Lutra canadensis.* Wildlife management. Mustelidae, Felidae.

39 **Anonymous.** 1978. Recovery charted for four species: Black-footed ferret tied to prairie dog management. Endangered Species Tech. Bull. 3(7):1,6.—Oklahoma cooperating in recovery plan for *Mustela nigripes.* Wildlife management. Sciuridae, Mustelidae.

40 **Anonymous.** 1978. Rulemaking actions—December 1977: Two big-eared bats. Endangered Species Tech. Bull. 3(1):4-5.—*Plecotus townsendii ingens* proposed for endangered status. Wildlife management. Vespertilionidae.

41 **Anonymous.** 1978. Wildlife portrait: The buffalo. Outdoor Okla. 34(5):24.—Bison (*Bison bison*). Account for the layman. History, Distribution, Natural history, Behavior, Ecology, Morphology, Popular. Bovidae.

42 **Anonymous.** 1978. Wildlife portrait: The coyote. Outdoor Okla. 34(8):20.—*Canis latrans.* Account for the layman. Natural history, Behavior, Ecology, Morphology, Popular. Canidae.

43 **Anonymous.** 1978. Wildlife portrait: The mink. Outdoor Okla. 34(9):20.—*Mustela vison.* Account for the layman. Natural history, Behavior, Ecology, Morphology, Popular. Mustelidae.

44 **Anonymous.** 1978. Wildlife portrait: The opossum. Outdoor Okla. 34(2):20.—Virginia opossum (*Didelphis virginiana*). Account for the layman. History, Distribution, Natural history, Behavior, Ecology, Morphology, Popular. Didelphidae.

45 **Anonymous.** 1978. Wildlife portrait: The whitetail. Outdoor Okla. 34(3):20.—White-tailed deer (*Odocoileus virginianus*). Account for the layman. Natural history, Behavior, Ecology, Morphology, Popular. Cervidae.

46 **Anonymous.** 1979. Two bats protected as endangered. Endangered Species Tech. Bull. 4(12):10.—*Plecotus townsendii ingens* listed as endangered. Wildlife management. Vespertilionidae.

47 **Anonymous.** 1979. Wildlife portrait: The elk. Outdoor Okla. 35(10):24.—*Cervus elaphus.* Account for the layman. History, Distribution, Natural history, Behavior, Ecology, Morphology, Popular. Cervidae.

48 **Anonymous.** 1979. Wildlife portrait: The prairie dog. Outdoor Okla. 35(4):24.—Black-tailed prairie dog (*Cynomys ludovicianus*). Account for the layman. Distribution, Natural history, Behavior, Ecology, Morphology, Popular. Sciuridae.

49 **Anonymous.** 1979. Wildlife portrait: The pronghorn. Outdoor Okla. 35(9):20.—*Antilocapra americana.* Account for the layman. History, Distribution, Natural history, Behavior, Ecology, Morphology, Popular. Antilocapridae.

50 **Anonymous.** 1979. Wildlife portrait: The red fox. Outdoor Okla. 35(3):20.—*Vulpes vulpes.* Account for the layman. Distribution, Natural history, Behavior, Ecology, Morphology, Popular. Canidae.

51 **Anonymous.** 1979. Wildlife portrait: Tree squirrels. Outdoor Okla. 35(6):20.—Gray squirrel (*Sciurus carolinensis*) and fox squirrel (*S. niger*). Account for the layman. Distribution, Natural history, Behavior, Ecology, Morphology, Popular. Sciuridae.

52 **Anonymous.** 1980. Investigations of big game species. Okla. Dep. Wildl. Cons. Job 01, Study 01, W-80-R-19. 74 pp.—Includes population and/or harvest data on elk (*Cervus elaphus*), pronghorn (*Antilocapra americana*), and white-tailed deer (*Odocoileus virgin-*

ianus). Emphasis on deer. Distribution, Populations, Wildlife management. Cervidae, Antilocapridae.

53 **Anonymous.** 1980. Wildlife portrait: The beaver. Outdoor Okla. 36(9):56-57.—*Castor canadensis.* Account for the layman. Natural history, Behavior, Ecology, Morphology, Popular. Castoridae.

54 **Anonymous.** 1980. Wildlife portrait: The gopher. Outdoor Okla. 36(6):24.—Thirteen-lined ground squirrel (*Spermophilus tridecemlineatus*). Account for the layman. Distribution, Natural history, Behavior, Ecology, Morphology, Popular. Sciuridae.

55 **Anonymous.** 1980. Wildlife portrait: The gray fox. Outdoor Okla. 36(2):20.—*Urocyon cinereoargenteus.* Account for the layman. Distribution, Natural history, Behavior, Ecology, Morphology, Popular. Canidae.

56 **Anonymous.** 1980. Wildlife portrait: The muskrat. Outdoor Okla. 36(5):24.—*Ondatra zibethicus.* Account for the layman. Natural history, Behavior, Ecology, Morphology, Popular. Cricetidae.

57 **Anonymous.** 1980. Wildlife portrait: The porcupine. Outdoor Okla. 36(3):24.—*Erethizon dorsatum.* Account for the layman. Natural history, Behavior, Ecology, Morphology, Popular. Erethizontidae.

58 **Anonymous.** 1981. Wildlife portrait: The bobcat. Outdoor Okla. 37(1):56-57.—*Felis rufus.* Account for the layman. Distribution, Natural history, Behavior, Ecology, Morphology, Popular. Felidae.

59 **Anonymous.** 1981. Wildlife portrait: The opossum. Outdoor Okla. 37(2):56-57.—Virginia opossum (*Didelphis virginiana*). Account for the layman. History, Distribution, Natural history, Behavior, Ecology, Morphology, Popular. Didelphidae.

60 **Anonymous.** 1981. Wildlife portrait: The red fox. Outdoor Okla. 37(5):56-57.—*Vulpes vulpes.* Account for the layman. Distribution, Natural history, Behavior, Ecology, Morphology, Popular. Canidae.

61 **Anonymous.** 1981. Wildlife portrait: The white-tail. Outdoor Okla. 37(6):56-57.—White-tailed deer (*Odocoileus virginianus*). Account for the layman. Natural history, Behavior, Ecology, Morphology, Popular. Cervidae.

62 **Anonymous.** 1981. Wildlife portrait: The woodchuck. Outdoor Okla. 37(4):56-57.—*Marmota monax.* Account for the layman. Distribution, Natural history, Behavior, Ecology, Morphology, Popular. Sciuridae.

63 **Anonymous.** 1982. Wildlife portrait: The raccoon. Outdoor Okla. 38(5):56-57.—*Procyon lotor.* Account for the layman. Natural history, Behavior, Ecology, Morphology, Popular. Procyonidae.

64 **Anonymous.** 1982. Wildlife portrait: The skunk. Outdoor Okla. 38(4):48-49.—Striped Skunk (*Mephitis mephitis*). Account for the layman. Natural history, Behavior, Ecology, Morphology, Popular. Mustelidae.

65 **Anonymous.** 1983. Wildlife portrait: The gray squirrel. Outdoor Okla. 39(1):56-57.—*Sciurus carolinensis.* Account for the layman. Natural history, Behavior, Ecology, Morphology, Popular. Sciuridae.

66 **Anonymous.** 1983. Wildlife portrait: The jackrabbit. Outdoor Okla. 39(4):56-57.—Black-tailed jackrabbit (*Lepus californicus*). Account for the layman. Distribution, Natural history, Behavior, Ecology, Morphology, Popular. Leporidae.

67 **Anonymous.** 1983. Wildlife portrait: The prairie dog. Outdoor Okla. 39(5):56-57.—Black-tailed prairie dog (*Cynomys ludovicianus*). Account for the layman. History, Natural history, Behavior, Ecology, Morphology, Popular. Sciuridae.

68 **Anonymous.** 1984. Wildlife portrait: The armadillo. Outdoor Okla. 40(2):56-57.—Nine-banded armadillo (*Dasypus novemcinctus*). Account for the layman. History, Natural history, Behavior, Ecology, Morphology, Popular. Dasypodidae.

69 **Anonymous.** 1984. Wildlife portrait: The muskrat. Outdoor Okla. 40(1):56-57.—*Ondatra zibethicus.* Account for the layman. Distribution, Natural history, Behavior, Ecology, Morphology, Popular. Cricetidae.

70 **Anonymous.** 1984. Wildlife portrait: The pronghorn. Outdoor Okla. 40(5):56-57.—*Antilocapra americana.* Account for the layman. History, Distribution, Natural history, Behavior, Ecology, Morphology, Popular. Antilocapridae.

71 **Anonymous.** 1985. Wildlife portrait: The eastern chipmunk. Outdoor Okla. 41(3):56-57.—*Tamias striatus.* Account for the layman. Natural history, Behavior, Ecology, Morphology, Popular. Sciuridae.

72 **Anonymous.** 1985. Wildlife portrait: The elk. Outdoor Okla. 41(5):56-57.—*Cervus elaphus.* Account for the layman. History, Distribution, Natural history, Behavior, Ecology, Morphology, Wildlife management, Popular. Cervidae.

73 **Anonymous.** 1985. Wildlife portrait: The weasel. Outdoor Okla. 41(1):56-57.—Long-tailed weasel (*Mustela frenata*). Account for the layman. History, Distribution, Natural history, Behavior, Ecology, Morphology, Popular. Mustelidae.

74 **Anonymous.** 1986. Wildlife portrait: The flying squirrel. Outdoor Okla. 42(5):56-57.—Southern flying squirrel (*Glaucomys volans*). Account for the layman. Distribution, Natural history, Behavior, Ecology, Morphology, Popular. Sciuridae.

75 **Anonymous.** 1986. Wildlife portrait: The opossum. Outdoor Okla. 42(2):56-57.—Virginia opossum (*Didelphis virginiana*). Account for the layman. History, Distribution, Natural history, Behavior, Ecology, Morphology, Popular. Didelphidae.

76 **Anonymous.** 1986. Wildlife portrait: The 13-lined ground squirrel. Outdoor Okla. 42(4):56-57.—*Spermophilus tridecemlineatus.* Account for the layman. Distribution, Natural history, Behavior, Ecology, Morphology, Popular. Sciuridae.

77 **Anthony, A.** 1953. Physiology of reproduction in the male prairie dog. Ph.D. Dissertation, Univ. Chicago, Chicago. 66 pp.—Laboratory colony of *Cynomys ludovicianus* from Comanche County. Natural history, Physiology. Sciuridae.

78 **Anthony, A.** 1953. Seasonal reproductive cycle in the normal and experimentally treated male prairie dog, *Cynomys ludovicianus.* J. Morphol. 93:331-369.—*Cynomys ludovicianus* specimens obtained from Wichita Mountains Wildlife Refuge, Comanche County. Natural history, Morphology, Physiology. Sciuridae.

79 **Anthony, A.** 1955. Behavior patterns in a laboratory colony of prairie dogs, *Cynomys ludovicianus.* J. Mammal. 36:69-78.—Laboratory colony of *Cynomys ludovicianus* from Comanche County. Natural history, Behavior, Physiology. Sciuridae.

80 **Anthony, A., and D. Foreman.** 1951. Observations on the reproductive cycle of the black-tailed prairie dog (*Cynomys ludovicianus*). Physiol. Zool. 24:242-248.—Specimens from Comanche County. Breeding is in late January or early February. Testicular activity begins in October; ovarian activity begins in December. Young arrive in March after 30 to 35 days gestation. Average litter size was 4.6. Natural history, Physiology. Sciuridae.

81 **Anthony, H. E.** 1928. Field book of North American mammals. G. P. Putnam's Sons, New York. 674 pp.—Field guide to mammals north of

Mexico. Systematics, Distribution, Natural history, Morphology, Bibliographies. Mammalia.

82 **Anthony, H. E.** 1937. Animals of America. Mammals of America. Garden City Publishing Co., Garden City, New York. 335 pp.—Systematics, History, Distribution, Natural history, Behavior, Ecology, Morphology, Wildlife management, Popular. Mammalia.

83 **Armstrong, D. M.** 1972. Distribution of mammals in Colorado. Univ. Kans. Mus. Nat. Hist. Monogr. 3:1-415.—Records of mammals along northwestern Oklahoma border. Includes keys. Systematics, Paleontology, History, Distribution, Natural history, Behavior, Ecology, Parasites, Morphology, Techniques, Economics, Wildlife management, Bibliographies. Mammalia.

84 **Armstrong, D. M., J. R. Choate, and J. K. Jones, Jr.** 1986. Distributional patterns of mammals in the plains states. Occas. Pap. Mus. Tex. Tech Univ. 105:1-27.—Analysis of distribution patterns of 119 mammal species found in North and South Dakota, Nebraska, Kansas, and Oklahoma. Superimposed distribution boundaries for each species were examined. Also included evaluation of habitats and purported biogeographic origin. History, Distribution, Ecology. Mammalia.

85 **Armstrong, D. M., and J. K. Jones, Jr.** 1972. *Notiosorex crawfordi.* Mamm. Species 17:1-5.—Distribution of *Notiosorex crawfordi* includes Oklahoma. Summary of information in literature. Systematics, Paleontology, Distribution, Natural history, Behavior, Ecology, Parasites, Genetics, Morphology, Physiology, Bibliographies. Soricidae.

Armstrong, D. M., *see* 642.

86 **Arnold, M. L., L. W. Robbins, R. K. Chesser, and J. C. Patton.** 1983. Phylogenetic relationships among six species of *Reithrodontomys.* J. Mammal. 64:128-132.—Study includes specimens of *Reithrodontomys fulvescens* from Murray County. Systematics, Genetics. Cricetidae.

87 **Arvey, M. D., and B. P. Glass.** 1950. The black-footed ferret in Oklahoma. J. Mammal. 31:460.—Specimen of *Mustela nigripes* in University of Oklahoma Stovall Museum [now Okla. Mus. Nat. Hist.] taken near Norman. Distribution. Mustelidae.

88 **Avise, J. C., M. H. Smith, and R. K. Selander.** 1974. Biochemical polymorphism and systematics in the genus *Peromyscus.* VI. The *boylii* species group. J. Mammal. 55:751-763.—Electrophoretic analysis of *Peromyscus boylii* subspecies and *P. attwateri,* including Oklahoma specimens. Systematics. Cricetidae.

Bahr, R. J., *see* 607.

89 **Bailey, V.** 1895. The pocket gophers of the United States. U.S. Dep. Agric. Bull. 5:1-47.—*Geomys breviceps* in eastern Oklahoma, *G. lutescens* in western Oklahoma, and *Cratogeomys castanops* in western Cimarron County. Systematics, Distribution, Natural history, Behavior, Ecology, Techniques, Economics. Geomyidae.

90 **Bailey, V.** 1898. Descriptions of eleven new species and subspecies of voles. Proc. Biol. Soc. Wash. 12:85-90.—Type locality of *Microtus pinetorum nemoralis* in Adair County. Systematics, Distribution, Morphology. Cricetidae.

91 **Bailey, V.** 1900. Revision of American voles of the genus *Microtus*. N. Am. Fauna 17:1-88.—Type locality of *Microtus nemoralis* (=*pinetorum*) at Stilwell, Indian Territory. Also *Microtus austerus* (=*ochrogaster*) reported from Oklahoma. Systematics, Distribution, Natural history, Behavior, Ecology, Morphology, Economics. Cricetidae.

92 **Bailey, V.** 1905. Biological survey of Texas. N. Am. Fauna 25:1-222.—Primarily a survey of mammals. Oklahoma distribution included in all accounts. Distribution, Natural history, Behavior, Ecology, Morphology. Mammalia.

93 **Bailey, V.** 1931. Mammals of New Mexico. N. Am. Fauna 53:1-412.—History, Distribution, Natural history, Behavior, Ecology, Economics, Wildlife management. Mammalia.

Baird, D. D., *see* 144.

94 **Baird, S. F.** 1857. Reports of explorations and surveys to ascertain the most practicable and economical route for a railroad. Vol. 8, Part 1, General report upon the zoology of the several Pacific Railroad routes. Part 1, Mammals. U.S. Senate, Washington, D. C. 757 pp.—Includes discussion of the mammals of Oklahoma and Indian territories. Systematics, History, Distribution, Natural history, Morphology, Bibliographies. Mammalia.

95 **Baird, S. F.** 1859. The mammals of North America. J. B. Lippincott and Co., Philadelphia. 764 pp.—Includes a few records from Oklahoma. Systematics, History, Distribution, Natural history, Morphology, Bibliographies. Mammalia.

96 **Baird, S. F.** 1974. Mammals of North America. Arno Press, New York. 764 pp.—Reprint of the 1859 edition published by Lippincott,

Philadelphia. Systematics, History, Distribution, Natural history, Morphology, Bibliographies. Mammalia.

97 **Baker, R. H., and B. P. Glass.** 1951. The taxonomic status of the pocket gophers, *Geomys bursarius* and *Geomys breviceps.* Proc. Biol. Soc. Wash. 64:55-58.—*Geomys breviceps* synonomized with *G. bursarius.* Specimens from Cleveland and Pottawatomie counties. Systematics, Distribution, Morphology. Geomyidae.

98 **Baker, R. J.** 1965. Systematics and variation of *Myotis subulatus* (Say). M.S. Thesis, Oklahoma State Univ., Stillwater. 48 pp.—Distribution of *Myotis l. leibi* (=*l. leibii*) includes southern and eastern Oklahoma. *Myotis subulatus* rejected as acceptable name. Systematics, Distribution, Morphology. Vespertilionidae.

99 **Baker, R. J., and J. T. Mascarello.** 1969. Karyotypic analyses of the genus *Neotoma* (Cricetidae, Rodentia). Cytogenetics (Basel) 8:187-198.—*Neotoma floridana, N. micropus, N. albigula, N. mexicana,* and other *Neotoma* species. Payne County, Oklahoma, as well as Texas, Arizona, California, Wyoming, and Mexico. Six species had diploid numbers of 52. Diploid numbers of 38, 54, and 56 also found. Extensive polymorphism in *N. lepida* and *N. micropus.* Chromosomal variation mainly in fundamental number and not diploid number. Genetics. Cricetidae.

100 **Baker, R. J., L. W. Robbins, F. B. Stangl, Jr., and E. C. Birney.** 1983. Chromosomal evidence for a major subdivision in *Peromyscus leucopus.* J. Mammal. 64:356-359.—Two cytotypes of *Peromyscus leucopus* differ in three chromosomes, with hybrid zone established in central Oklahoma. Systematics, Genetics. Cricetidae.

101 **Baker, R. J., and D. L. Spencer.** 1965. Late fall reproduction in the desert shrew. J. Mammal. 46:330.—Specimens of *Notiosorex crawfordi* from Harmon County. Distribution, Natural history. Soricidae.

Baker, R. J., *see* 271, 392, 603, 756, 1023.

102 **Baldwin, J.** 1969. The Lawrence site, Nw-6, a non-ceramic site in Nowata County, Oklahoma. Bull. Okla. Anthropol. Soc. 18:67-118.—Workers recovered 2,760 bone fragments, but were not able to identify 76.5%. Of those identified, 17.4% represented white-tailed deer (*Odocoileus virginianus*). Other mammals also present. Paleontology. Cervidae, Mammalia.

103 **Bangs, O.** 1896. Some new mammals from Indian Territory and Missouri. Proc. Biol. Soc. Wash. 10:135-138.—Type localities of *Lepus sylvaticus* (=*Sylvilagus floridanus*) *alacer, Peromyscus bellus*

(=*attwateri*), *Tamias striatus venustus,* and *Scalops texanus* (=*Scalopus aquaticus*) *aereus* in Adair County. Systematics, Distribution, Morphology. Talpidae, Leporidae, Sciuridae, Cricetidae.

104 **Barbour, R. W., and W. H. Davis.** 1969. Bats of America. Univ. Press Kentucky, Lexington. 286 pp.—Dichotomous key, descriptions, and range maps showing approximate distributions in Oklahoma. Paleontology, Distribution, Natural history, Ecology, Populations, Parasites, Morphology, Physiology, Economics, Wildlife management, Bibliographies. Vespertilionidae, Molossidae.

Barclay, B. D., *see* 233.

105 **Barclay, J. S.** 1980. Impact of stream alterations on riparian communities in southcentral Oklahoma. U.S. Fish Wildl. Serv. Program FWS/OBS 80/17.—Small and large mammal communities sampled on Rush Creek (Garvin County), Wildhorse Creek (Garvin and Stephens counties), Lake Creek and Fort Cobb Reservoir (Caddo County), and Cobb Creek (Caddo and Washita counties). Distribution, Ecology, Techniques, Wildlife management. Mammalia.

Barclay, R. M. R., *see* 353.

106 **Bareis, C. J.** 1955. Archaeological and paleontological survey trip to Alva, Oklahoma. Okla. Anthropol. Soc. Newsl. 3(7):3.—Bones of mammoth and/or mastodon found at two sites in Woods County. Paleontology. Mastodontidae, Elephantidae.

107 **Barkley, F. A., and C. C. Smith.** 1934. A preliminary study of buffalo wallows in the vicinity of Norman, Oklahoma. Proc. Okla. Acad. Sci. 14:47-52.—Describe physical and vegetational characteristics of wallows made by bison (*Bison bison*). Wallows provide local habitats for prairie plant species adapted to mesic conditions. Ecology. Bovidae.

Barlow, R., *see* 15.

108 **Barr, T. P.** 1966. The Pruitt site: A Late Plains Woodland manifestation in Murray County, Oklahoma. Okla. River Basin Surv. Archaeol. Site Rep. 5:1-126.—Mammals were major source of food for people at prehistoric site in Murray County. Paleontology. Leporidae, Sciuridae, Castoridae, Cricetidae, Procyonidae, Cervidae.

Barrett, S. D., *see* 1084.

Barron, S. J., *see* 1183.

109 **Bartush, W. S.** 1978. Mortality of white-tailed deer fawns in the Wichita Mountains, Comanche County, Oklahoma, Part II. M.S.

Thesis, Oklahoma State Univ., Stillwater. 161 pp.—*Odocoileus virginianus* fawns in Comanche County suffered high mortality. Activity patterns of does and fawns also studied. Natural history, Behavior, Ecology, Populations, Techniques, Wildlife management. Cervidae.

110 **Bartush, W. S., and G. W. Garner.** 1979. Physical characteristics of white-tailed deer fawns in southwestern Oklahoma. Proc. Annu. Conf. Southeast. Assoc. Fish Wildl. Agencies 33:250-258.—Growth curves determined for young *Odocoileus virginianus* in Comanche County. Natural history, Morphology. Cervidae.

111 **Bartush, W. S., and J. C. Lewis.** 1978. Behavior of whitetail does and fawns during the parturition period. Proc. Annu. Conf. Southeast. Assoc. Fish Wildl. Agencies 32:246-255.—*Odocoileus virginianus* does separated themselves from conspecifics prior to parturition. Natural history, Behavior, Populations. Cervidae.

112 **Bartush, W. S., and J. C. Lewis.** 1981. Mortality of white-tailed deer fawns in the Wichita Mountains. Proc. Okla. Acad. Sci. 61:23-27.—In Comanche County 90% of *Odocoileus virginianus* fawns died within three months, most as a result of coyote (*Canis latrans*) predation. Natural history, Behavior, Populations, Wildlife management. Canidae, Felidae, Cervidae.

Bartush, W. S., *see* 728.

113 **Base, D. L.** 1986. Evaluation of experimental reintroduction of river otters into Oklahoma 1984-1986. Nongame Wildlife Program, Oklahoma Department of Wildlife Conservation, Oklahoma City. 40 pp.—Results of introductions of river otters (*Lutra canadensis*) from coastal Louisiana into southeastern Oklahoma. Natural history, Behavior, Ecology, Populations, Wildlife management. Mustelidae.

114 **Baskin, J. A.** 1982. Tertiary Procyoninae (Mammalia: Carnivora) of North America. J. Vertebr. Paleontol. 2:71-93.—New Pliocene species (*Arctonasua fricki*) from Texas County. Paleontology. Procyonidae.

115 **Baugh, S. T.** 1986. Late Prehistoric bison distributions in Oklahoma. Plains Anthropol. 114:93-96.—Records of *Bison* from 500 to 1300 A.D. Paleontology, Distribution. Bovidae.

116 **Baumgardner, G. D., and D. J. Schmidly.** 1981. Systematics of the southern races of two species of kangaroo rats (*Dipodomys compactus* and *D. ordii*). Occas. Pap. Mus. Tex. Tech Univ. 73:1-27.—Study includes Cleveland County specimens of *Dipodomys ordii oklahomae.* Systematics, Distribution, Morphology. Heteromyidae.

Baumgardner, G. D., *see* 654.

117 **Baumgartner, A. M., and F. M. Baumgartner.** 1944. Hawks and owls in Oklahoma 1939-1942: Food habits and population changes. Wilson Bull. 56:209-215.—Ten mammal species comprised most of the food of six predatory bird species in Payne County. Natural history, Ecology, Populations. Soricidae, Talpidae, Leporidae, Heteromyidae, Cricetidae, Muridae.

118 **Baumgartner, F. M.** 1948. A preliminary study of the effects of certain insecticides upon wildlife in north central Oklahoma. Proc. Okla. Acad. Sci. 29:6-10.—Study near Stillwater includes *Sciurus niger.* Ecology. Sciuridae.

Baumgartner, F. M., *see* 117.

119 **Bean, C.** 1983. Mountain lion sightings growing with deer population. Daily Oklahoman, December 12. Oklahoma City.—Reports of recent specimens and sightings in Oklahoma of mountain lions (*Felis concolor*). Increased numbers seem to be associated with increases in deer numbers. Distribution, Popular. Felidae, Cervidae.

Bearden, S., *see* 236.

120 **Beck, M. L., M. L. Kennedy, and C. J. Biggers.** 1981. Genic variation in Ord's kangaroo rat *Dipodomys ordii* in Oklahoma. Am. Midl. Nat. 105:77-83.—Polyacrilamide electrophoresis supports conspecific status of *Dipodomys ordii richardsoni* and *D. o. oklahomae.* Systematics, Distribution, Genetics. Heteromyidae.

Beck, M. L., *see* 652.

121 **Beckett, G. E.** 1963. Selected skeletal development in Mexican freetailed bats, *Tadarida brasiliensis mexicana* (Saussure). M.S. Thesis, Oklahoma State Univ., Stillwater. 37 pp.—Study assessed skeletal growth and ossification in *Tadarida brasiliensis* from Woodward County. Natural history, Morphology, Techniques. Molossidae.

122 **Bee, J. W., G. E. Glass, R. S. Hoffmann, and R. R. Patterson.** 1981. Mammals in Kansas. Univ. Kans. Mus. Nat. Hist. Public Education Ser. 7:1-300.—Updated book including records of mammals along northern Oklahoma border. Distribution, Natural history, Ecology, Behavior. Mammalia.

123 **Bekoff, M.** 1977. *Canis latrans.* Mamm. Species 79:1-9.—Distribution of coyote (*Canis latrans*) includes Oklahoma. Summary of information in literature. Systematics, Paleontology, Distribution, Natural history, Behavior, Ecology, Populations, Parasites, Genetics, Morphol-

ogy, Physiology, Techniques, Economics, Wildlife management, Bibliographies. Canidae.

124 **Bekoff, M., and R. Jamieson.** 1975. Physical development in coyotes (*Canis latrans*), with a comparison of other canids. J. Mammal. 56:685-692.—Post-natal development. Some specimens of Oklahoma origin. Natural history, Morphology. Canidae.

125 **Bell, R. E.** 1969. Oklahoma archaeology: An annotated bibliography. Univ. Oklahoma Press, Norman. 99 pp.—Annotated bibliography of scientific literature on Oklahoma archaelogy. Bibliographies. Mammalia.

126 **Bell, R. E.** 1971. Bison scapula skin-dressing tools? Plains Anthropol. 16:125-127.—Requested information concerning bison (*Bison bison*) scapula tools like those associated with Washita River focus. Paleontology. Bovidae.

127 **Bell, R. E.** 1978. Oklahoma archeology: An annotated bibliography. Second edition. Univ. Oklahoma Press, Norman. 155 pp.—Update of 1969 bibliography. Author added 291 references to bring total to 786. Bibliographies. Mammalia.

128 **Best, T. L.** 1971. Notes on the distribution and ecology of five eastern New Mexico mammals. Southwest. Nat. 16:210-211.—Suggested that Cimarron River may be a route of dispersal for the white-tailed deer (*Odocoileus virginiana texana*), perhaps into extreme northwestern part of Oklahoma Panhandle. Distribution. Cervidae.

129 **Best, T. L., and B. Hoditschek.** 1982. Analysis of cheek pouch contents of Ord's kangaroo rat (*Dipodomys ordii*). Southwest. Nat. 27:117-119.—*Dipodomys ordii oklahomae* cheek pouches contained mostly seeds; no animal material. Natural history, Ecology. Heteromyidae.

130 **Best, T. L., B. Hoditschek, and H. H. Thomas.** 1981. Foods of coyotes (*Canis latrans*) in Oklahoma. Southwest. Nat. 26:67-69.—Coyotes (*Canis latrans*) in Cleveland, Grady, and McClain counties eat grasses, fruits, insects, birds, and mammals. Cattle (probably carrion) were predominant food of central Oklahoma coyotes. Ecology. Canidae, Mammalia.

131 **Best, T. L., and M. L. Kennedy.** 1972. The porcupine (*Erethizon dorsatum* Linnaeus) in the Texas Panhandle and adjacent New Mexico and Oklahoma. Tex. J. Sci. 24:351.—Cimarron, Comanche, Harper, Jackson, Ellis, and Woods counties, Oklahoma, as well as Texas and New Mexico. Additional occurrence records. Specimens at Univer-

sity of Oklahoma Stovall Museum [now Okla. Mus. Nat. Hist.] and Eastern New Mexico University Natural History Museum. Distribution. Erethizontidae.

132 **Best, T. L., and G. D. Schnell.** 1974. Bacular variation in kangaroo rats (genus *Dipodomys*). Am. Midl. Nat. 91:257-270.—Study includes *Dipodomys ordii* from Oklahoma. Systematics, Morphology. Heteromyidae.

Best, T. L., *see* 150, 584, 652, 902.

Biggers, C. J., *see* 120.

133 **Bigham, S. R.** 1965. Breeding season of the cottontail rabbit in north-central Oklahoma. Proc. Okla. Acad. Sci. 46:217-219.—*Sylvilagus floridanus alacer* in Payne and Noble counties. Breeding season from mid-February to early September. Natural history. Leporidae.

134 **Bigham, S. R.** 1965. Reproduction of the wild cottontail rabbit in north central Oklahoma. M.S. Thesis, Oklahoma State Univ., Stillwater. 42 pp.—Reproductive and population characteristics of *Sylvilagus floridanus* studied in Payne and Noble counties. Natural history, Ecology, Populations, Physiology, Wildlife management. Leporidae.

135 **Bigham, S. R.** 1966. Identification of the habitat of the cottontail rabbit in central Oklahoma: Identification of vegetation used for shelter as indicated by the presence of forms, and by cover used for retreat. Okla. Dep. Wildl. Cons. Job 02, W-089-R-01. 9 pp.—Study of habitat utilized by *Sylvilagus floridanus.* Natural history, Ecology, Populations, Techniques. Leporidae.

136 **Bigham, S. R.** 1966. Identification of the habitat of the cottontail rabbit in central Oklahoma: Identification of plants used for food as indicated by examination of stomach contents and food residues in fecal pellets. Okla. Dep. Wildl. Cons. Job 01, W-089-R-01. 3 pp.—Study of habitat and foods of *Sylvilagus floridanus.* Natural history, Ecology, Techniques. Leporidae.

137 **Bigham, S. R.** 1970. The comparative utility of plant life-form and other vegetational characteristics in evaluating the habitat of the cottontail. Ph.D. Dissertation, Oklahoma State Univ., Stillwater. 69 pp. —Plant life-form, stem density, and canopy density correlated with use by *Sylvilagus floridanus* in Payne County. Natural history, Behavior, Ecology, Wildlife management. Leporidae.

138 **Bigham, S. R., J. L. Hepworth, and R. P. Martin.** 1964. A casuality count of wildlife following a fire. Proc. Okla. Acad. Sci. 45:47-50.—Rodents and rabbits seen fleeing fire, but no reported mammal casualties. Ecology. Mammalia.

139 **Bird, R. D.** 1930. Cottontail rabbits are insectivorous. J. Mammal. 11:240.—*Sylvilagus floridanus* ate silk moth cocoons. Natural history, Behavior. Leporidae.

140 **Bird, R. D.** 1933. A three-horned wapiti (*Cervus canadensis canadensis*). J. Mammal. 14:164-166.—*Cervus canadensis* (=*elaphus*) with two left antlers, Wichita Mountains Refuge, Comanche County. Morphology. Cervidae.

141 **Birney, E. C.** 1973. Systematics of three species of woodrats (genus *Neotoma*) in central North America. Univ. Kans. Mus. Nat. Hist. Misc. Publ. 58:1-173.—Study includes *Neotoma micropus canescens* and *N. floridana attwateri* from Woods and Major counties. Systematics, History, Distribution, Parasites, Genetics, Morphology. Cricetidae.

142 **Birney, E. C.** 1974. Dynamics of small mammal populations at the Cottonwood and Osage sites, 1972. U.S. IBP (Int. Biol. Program) Grassland Biome Tech. Rep. 257:1-87. Colorado State Univ., Fort Collins.—Small mammals censused in three seasons at Osage County site. Ecology, Populations, Techniques. Soricidae, Sciuridae, Heteromyidae, Cricetidae, Muridae.

143 **Birney, E. C.** 1976. An assessment of relationships and effects of interbreeding among woodrats of the *Neotoma floridana* species-group. J. Mammal. 57:103-132.—Study includes *Neotoma floridana, N. micropus,* and *N. albigula* from Oklahoma. Systematics, Genetics, Morphology. Cricetidae.

144 **Birney, E. C., W. E. Grant, and D. D. Baird.** 1976. Importance of vegetative cover to cycles of *Microtus* populations. Ecology 57:1043-1051.—*Microtus ochrogaster* in Osage County. Site-dependent threshold level of vegetative cover is necessary for a population of *Microtus* to increase enough to undergo a multiyear cyle. Distribution, Natural history, Ecology, Populations. Cricetidae, Mammalia.

Birney, E. C., *see* 100, 449, 450, 585, 913.

Bissell, S. J., *see* 815.

145 **Bissonette, J. A., and O. E. Maughan.** 1978. Southeastern Oklahoma coal investigations: Endangered species. Proj. No. 14-16-002-77-080

to U.S. Fish and Wildlife Service. Okla. Coop. Wildl. Fish. Res. Units, Oklahoma State Univ., Stillwater. 101 pp.—Study in Atoka, Coal, Haskell, Latimer, LeFlore, and Pittsburg counties indicates presence of *Lutra canadensis, Lynx* (=*Felis*) *rufus, Felis concolor,* and possibly *Myotis sodalis.* Distribution, Populations, Wildlife management. Vespertilionidae, Canidae, Ursidae, Mustelidae, Felidae.

Bissonette, J. A., *see* 844, 845, 846.

Black, C. C., *see* 674.

146 **Black, J. D.** 1935. A new race of *Peromyscus maniculatus* from Arkansas. J. Mammal. 16:144-145.—*Peromyscus maniculatus ozarkiarum.* Includes mention of Oklahoma distribution. Systematics, Distribution. Cricetidae.

147 **Black, J. D.** 1936. Mammals of northwestern Arkansas. J. Mammal. 17:29-35.—Oklahoma localities listed for *Peromyscus maniculatus* and *P. attwateri.* Distribution. Cricetidae, Mammalia.

148 **Black, J. D.** 1937. Mammals of Kansas. Kans. State Board Agric. Biennial Rep. 30:116-217.—Includes key for all Kansas mammals, most of which occur in Oklahoma. Systematics, Distribution, Natural history, Behavior, Ecology, Morphology, Techniques, Wildlife management. Mammalia.

149 **Black, J. H.** 1976. Escape behavior of the southern plains skink. Herpetol. Rev. 7:111.—Prairie skink (*Eumeces septentrionalis*) used muskrat (*Ondatra zibethicus*) burrow as a retreat. Natural History. Cricetidae.

150 **Black, J. H., and T. L. Best.** 1972. Remains of a gray wolf (*Canis lupus*) from northwestern Oklahoma. Proc. Okla. Acad. Sci. 52:120.—Woodward County. History, Distribution, Ecology, Morphology. Canidae.

151 **Black, J. H., R. Hunsicker, A. P. Blair, and T. Hunkapillar.** 1977. The eastern chipmunk in central Oklahoma with validation of Howell's 1929 Red Fork record. Proc. Okla. Acad. Sci. 57:163-164.—*Tamias striatus.* Distribution. Sciuridae.

152 **Blackwood, M. A.** 1981. Factors determining the distribution of *Plecotus* in North America. M.S. Thesis, Oklahoma State Univ., Stillwater. 59 pp.—Literature review and discussion of *Plecotus* species and factors influencing their distributions in North America north of Mexico. Oklahoma species include Townsend's big-eared bat (*P.*

townsendii) and Rafinesque's big-eared bat (*P. rafinesquii*). Distribution, Natural history, Ecology, Bibliographies. Vespertilionidae.

Blair, A. P., *see* 151, 164.

153 **Blair, K. B.** 1978. The effects of channelization and impoundment on riparian small mammal communities in south central Oklahoma. M.S. Thesis, Oklahoma State Univ., Stillwater. 46 pp.—Study in Caddo, Grady, and Carter counties shows unchannelized stream bank communities to have higher small-mammal species diversities. Ecology, Populations. Soricidae, Talpidae, Geomyidae, Heteromyidae, Cricetidae, Muridae.

154 **Blair, W. F.** 1936. The nine-banded armadillo in northeastern Oklahoma. J. Mammal. 17:293-294.—*Dasypus novemcinctus.* Distribution. Dasypodidae.

155 **Blair, W. F.** 1937. The burrows and food of the prairie pocket mouse. J. Mammal. 18:188-191.—*Perognathus* (=*Chaetodipus*) *hispidus.* Behavior, Ecology. Heteromyidae.

156 **Blair, W. F.** 1938. Ecological relationships of the mammals of the Bird Creek region, northeastern Oklahoma. Am. Midl. Nat. 20:473-526.—Intensive study of ecological associations of mammals in Tulsa and Rogers counties. Distribution, Natural history, Ecology, Populations. Mammalia.

157 **Blair, W. F.** 1938. A new race of the eastern cottontail from the Texas Panhandle. Occas. Pap. Mus. Zool. Univ. Mich. 380:1-3.—Specimens of *Sylvilagus floridanus* examined from Harper, Rogers, and Pottawatomie counties. Systematics, Distribution. Leporidae.

158 **Blair, W. F.** 1939. Faunal relationships and geographic distribution of mammals in Oklahoma. Am. Midl. Nat. 22:85-133.—Annotated mammal list with Oklahoma records. Biotic districts. Distribution, Ecology. Mammalia.

159 **Blair, W. F.** 1939. New mammals from Texas and Oklahoma, with remarks on the status of *Thomomys texensis baileyi.* Occas. Pap. Mus. Zool. Univ. Mich. 403:1-7.—Type locality of *Neotoma floridana osagensis.* Systematics. Cricetidae.

160 **Blair, W. F.** 1939. Some observed effects of stream-valley flooding on mammalian populations in eastern Oklahoma. J. Mammal. 20:304-306.—Flooding reduced populations of small terrestrial mammals. Ecology, Populations. Soricidae, Talpidae, Leporidae, Geomyidae, Cricetidae.

161 **Blair, W. F.** 1943. Biological and morphological distinctness of a previously undescribed species of the *Peromyscus truei* group from Texas. Contr. Lab. Vertebr. Biol. Univ. Mich. 24:1-8.—Specimens of *Peromyscus comanche* examined from Cimarron County. Systematics, Distribution, Morphology. Cricetidae.

162 **Blair, W. F.** 1949. Extensions of the known range of three species of Texas mammals. J. Mammal. 30:201-202.—*Dipodomys elator.* Includes Oklahoma localities. Distribution, Ecology. Heteromyidae.

163 **Blair, W. F.** 1954. Mammals of the Mesquite Plains biotic district in Texas and Oklahoma, and speciation in the central grasslands. Tex. J. Sci. 6:235-264.—Jackson, Harmon, and Cotton counties, Oklahoma, and 12 Texas counties. Annotated list of 60 species of mammals. Ecological associations. Speciation through isolation of population fragments due to Pleistocene and post-Pleistocene species shifts. Systematics, Distribution, Ecology. Mammalia.

164 **Blair, W. F., A. P. Blair, P. Brodkorb, F. G. Cagle, and G. A. Moore.** 1968. Vertebrates of the United States. Second edition. McGraw-Hill Book Co., New York. 616 pp.—Keys to species of mammals. Distribution, Ecology, Morphology. Mammalia.

165 **Blair, W. F., and T. H. Hubbell.** 1938. The biotic districts of Oklahoma. Am. Midl. Nat. 20:425-454.—Brief discussion of ecological biogeography of the mammals of Oklahoma. Distribution, Ecology. Mammalia.

Blocker, H. D., *see* 913.

Blouin, E. F., *see* 680.

166 **Bobalik, S. J., and J. R. Galm.** 1978. Faunal remains from the Scott (34LF-11) and Wann (34LF-27) sites, Oklahoma. Pages 299-332 (Appendix C) *in* The cultural sequences at the Scott (34LF-11) and Wann (34LF-27) sites and prehistory of the Wister Valley (J. R. Galm and P. Flynn, eds.). Univ. Okla. Archaeol. Res. and Manage. Center Res. Ser. No. 3. 350 pp.—White-tailed deer (*Odocoileus virginianus*) were most common at sites, with elk (*Cervus elaphus*) also found. Twelve other mammal species reported. Paleontology. Cervidae, Mammalia.

Bogan, A. E., *see* 863.

167 **Bogan, M., and P. Mehlhop.** 1983. Systematic relationships of gray wolves (*Canis lupus*) in southwestern North America. Occas. Pap.

Mus. Southwest. Biol. 1:1-21.—Study includes specimens of *Canis lupus nubilus* from Oklahoma. Systematics, Distribution. Canidae.

Boggess, E. K., *see* 241.

168 **Bohlin, R. G., and E. G. Zimmerman.** 1982. Genic differentiation of two chromosome races of the *Geomys bursarius* complex. J. Mammal. 63:218-228.—Electrophoretic study indicates that *Geomys bursarius breviceps* is a distinct species. Systematics, Distribution, Genetics. Geomyidae.

169 **Bole, B. P., Jr., and P. N. Moulthrop.** 1942. The Ohio Recent mammal collection in the Cleveland Museum of Natural History. Sci. Publ. Cleveland Mus. Nat. Hist. 5:83-181.—Oklahoma specimens of *Blarina brevicauda hulophaga* (=*h. hylophaga*) examined. Distribution. Soricidae.

170 **Bolte, J. R., J. A. Hair, and J. Fletcher.** 1970. White tailed deer mortality following tissue destruction induced by lone star ticks. J. Wildl. Manage. 34:546-552.—*Odocoileus virginianus* fawns suffered high mortality in eastern Oklahoma, primarily from lone star tick infestations. Natural history, Populations, Parasites. Cervidae.

171 **Bonaccorso, F. J., N. Smythe, and S. R. Humphrey.** 1976. Improved techniques for marking bats. J. Mammal. 57:181-182.—Punch-marking useful only in short-term studies. Suggestions for banding. Techniques. Vespertilionidae.

172 **Booth, E. S.** 1961. How to know the mammals. Second edition. Wm. C. Brown Co., Dubuque, Iowa. 203 pp.—Dichotomous keys, descriptions, and range maps showing approximate distributions in Oklahoma. Distribution, Natural history, Morphology, Techniques. Mammalia.

Bousman, C. B., *see* 236.

173 **Bowers, J. H.** 1974. Genetic compatability of *Peromyscus maniculatus* and *Peromyscus melanotis,* as indicated by breeding studies and morphometrics. J. Mammal. 55:720-737.—*Peromyscus maniculatus* and *P. melanotis* judged to be sibling species, not conspecifics. Systematics, Genetics, Morphology. Cricetidae.

174 **Boyd, K. M.** 1983. Zooarchaeological analyses. Pages 220-232 *in* Bug Hill: Excavation of a multicomponent midden mound in the Jackfork Valley, southeast Oklahoma. Part 1 (J. H. Altschul, ed.). Report for Tulsa District Corps of Engineers. New World Research, Inc., Pollack, Louisiana. 232 pp.—Faunal remains at this Holocene site in-

cluded over a million bones of which 26,000 examined. Paleontology. Didelphidae, Talpidae, Leporidae, Sciuridae, Castoridae, Cricetidae, Canidae, Procyonidae, Cervidae.

175 **Brabander, J. J.** 1977. Wildlife population and habitat evaluation on a northcentral Oklahoma site using Landsat-1 imagery. M.S. Thesis, Oklahoma State Univ., Stillwater. 145 pp.—Study in Payne County shows satellite-imagery mapping correlates with ground-based mapping of vegetation types. Mammal populations assessed in several habitat types. Ecology, Populations, Techniques, Wildlife management. Mammalia.

176 **Bridgwater, D. D.** 1966. Laboratory breeding, early growth, development and behavior of *Citellus tridecemlineatus* (Rodentia). Southwest. Nat. 11:325-337.—Most specimens of thirteen-lined ground squirrel, *Citellus* (=*Spermophilus*) *tridecemlineatus,* used in study were trapped near Oklahoma City. Age of occurrence of certain behavioral and developmental phenomena recorded. Post-natal growth curves included. Natural history, Behavior, Morphology. Sciuridae.

177 **Bridgwater, D. D., and D. F. Penney.** 1966. Predation by *Citellus tridecemlineatus* on other vertebrates. J. Mammal. 47:345-346.—Thirteen-lined ground squirrel, *Citellus* (=*Spermophilus*) *tridecemlineatus,* observed taking young eastern cottontails, *Sylvilagus floridanus,* as well as other vertebrates. Natural history. Leporidae, Sciuridae.

Brodkorb, P., *see* 164.

178 **Brooks, R. L., R. R. Drass, and F. E. Swenson.** 1985. Prehistoric farmers of the Washita River Valley: Settlement and subsistence patterns during the plains village period. Okla. Archeol. Surv., Okla. Archeol. Resource Surv. Rep. No. 23. 162 pp.—Faunal remains identified include bison (*Bison bison*), white-tailed deer (*Odocoileus virginianus*), eastern cottontails (*Sylvilagus floridana*), and squirrels (*Sciurus* sp.). Paleontology. Leporidae, Sciuridae, Cervidae.

Brown, A. V., *see* 1176.

179 **Brown, H. P., and W. D. Shepard.** 1979. Type specimens in the zoological collections of the Stovall Museum of Science and History. Southwest. Nat. 24:195-197.—Topotype of *Dipodomys ordii oklahomae* Trowbridge and Whitaker. Systematics. Heteromyidae.

180 **Brown, M.** 1975. The beaver in Oklahoma. Outdoor Okla. 31(1):12-14.—*Castor canadensis.* Account for the layman. History, Natural his-

tory, Behavior, Ecology, Morphology, Economics, Wildlife management, Popular. Castoridae.

181 **Brown, M.** 1975. Cougar. Outdoor Okla. 31(2):11-13.—*Felis concolor.* Account for the layman. Distribution, Natural history, Behavior, Ecology, Popular. Felidae.

182 **Brown, M.** 1975. The night creatures. Outdoor Okla. 31(3):5-7.—Account for the layman. Mentions bats and flying squirrels. Systematics, Behavior, Morphology, Popular. Vespertilionidae, Molossidae, Sciuridae.

183 **Bryles, M. C., G. C. Cozad, and A. Robinson.** 1969. Isolation of *Histoplasma capsulatum* from bats in Oklahoma. Am. J. Trop. Med. Hyg. 18:399-400.—*Histoplasma capsulatum* recovered from *Tadarida brasiliensis mexicana* in cave near Reed, Greer County. Parasites. Molossidae.

184 **Buchanan, G. D.** 1958. The current range of the armadillo *Dasypus novemcinctus mexicanus* in the United States. Tex. J. Sci. 10:349-351.—*Dasypus novemcinctus* had expanding distribution in Oklahoma. Distribution. Dasypodidae.

185 **Buchanan, G. D., and R. V. Talmage.** 1954. The geographical distribution of the armadillo in the United States. Tex. J. Sci. 6:142-150.—*Dasypus novemcinctus* distribution was expanding in Oklahoma and elsewhere. Distribution. Dasypodidae.

Buchanan, G. D., *see* 1070.

186 **Buechner, H. K.** 1950. Range ecology of the pronghorn on the Wichita Mountains Wildlife Refuge. Trans. N. Am. Wildl. Conf. 1950:627-644.—History and 1949 status of *Antilocapra americana* herd on Wichita Mountains Wildlife Refuge, Comanche County. History, Populations. Antilocapridae.

187 **Burnett, C. D.** 1983. Geographic and climatic correlates of morphological variation in *Eptesicus fuscus.* J. Mammal. 64:437-444.—Morphologic variation evaluated geographically with respect to environmental variation. Study included some Oklahoma specimens. Ecology, Morphology. Vespertilionidae.

188 **Burnett, C. D.** 1983. Geographic and secondary sexual variation in the morphology of *Eptesicus fuscus.* Ann. Carnegie Mus. 52:139-162. —Geographic and secondary sexual variation in big brown bats. Includes Oklahoma materials. Systematics, Distribution, Morphology. Vespertilionidae.

189 **Burnham, G. L.** 1953. A study of the helminth parasites of the pocket gophers of Woods, Alfalfa, Grant, and Marshall counties, Oklahoma. Proc. Okla. Acad. Sci. 34:59-61.—*Geomys bursarius* parasitized by *Hymenolepis dimunata, Monoecocestus anoplocphaloides, Cittotaenia perplexa, Mastophorus muris ascaroides, Litomosa filaria,* and *Ostertagia* sp. Parasites. Geomyidae.

190 **Burns, R. D.** 1960. Recent additions to the mammal collection of the Stovall Museum. Proc. Okla. Acad. Sci. 41:70-72.—List of specimens and localities. Note on *Bassariscus astutus.* Distribution, Ecology. Procyonidae, Mammalia.

Burns, R. D., *see* 1, 805.

191 **Burt, W. H.** 1945. The Bunker bat in Oklahoma. J. Mammal. 26:309. —*Antrozous bunkeri* (=*pallidus*) in Cimarron County. Distribution. Vespertilionidae.

192 **Burt, W. H., and R. P. Grossenheider.** 1952. A field guide to the mammals. Houghton Mifflin Co., Boston. 200 pp.—Distribution, Natural history, Behavior, Ecology, Morphology, Popular. Mammalia.

193 **Burt, W. H., and R. P. Grossenheider.** 1964. A field guide to the mammals. Second edition. Houghton Mifflin Co., Boston. 284 pp.—Distribution, Natural history, Behavior, Ecology, Morphology, Popular. Mammalia.

194 **Burton, R. J.** 1969. Bison find in Garvin County. Okla. Anthropol. Newsl. 17(4):6.—*Bison* remains found in northern Garvin County. Paleontology. Bovidae.

195 **Burton, R. J.** 1971. The archaeological view from the Harvey site, Sq-18. Okla. River Basin Surv. Archaeol. Site Rep. 21:1-98.—Prehistoric site in Sequoyah County yielded remains of 10 mammal species. Paleontology. Mammalia.

196 **Buscher, H. N.** 1975. *Raillietina* (*Raillietina*) *selfi* sp. n. (Cestoda: Davaineidae) from desert cottontail in Oklahoma with notes on the distribution of *Raillietina* from North American mammals. Proc. Okla. Acad. Sci. 55:103-107.—*Sylvilagus audubonii* host for *Raillietina* (*Raillietina*) *selfi.* Key for North American species of *Raillietina.* Distribution, Parasites. Leporidae, Sciuridae, Geomyidae, Heteromyidae, Cricetidae.

197 **Buscher, H. N., and J. D. Tyler.** 1975. Parasites of vertebrates inhabiting prairie-dog towns in Oklahoma, II. Helminthes. Proc. Okla. Acad. Sci. 55:108-111.—*Cynomys ludovicianus, Sylvilagus audubonii,*

and *Spermophilus tridecemlineatus.* Nine helminths including two trematodes and two nematodes. New host records. Host-parasite relationships. Ecology, Parasites. Leporidae, Sciuridae.

Buscher, H. N., *see* 1107.

Busher, P. E., *see* 633.

Bussjaeger, L. J., *see* 697.

198 **Butler, B., and B. C. Yates.** 1982. Animal exploitation in the Delaware Canyon area. Pages 117-156 *in* The Late Holocene prehistory of Delaware Canyon, Oklahoma (C. R. Ferring, ed.). Institute of Applied Sciences. North Texas State Univ., Denton. 331 pp.— Remains include a large number of items representing many mammal species. Paleontology. Mammalia.

199 **Butler, B. H.** 1978. Vertebrate faunal remains. Pages 49-78 *in* The prehistory and paleoenvironment of Hominy Creek Valley (D. O. Henry, ed.). Contributions in Archaeology, Univ. Tulsa, Oklahoma.— Includes summary of over 60,000 bone pieces found at northeastern Oklahoma site. Rodents not treated in detail. Paleontology. Mammalia.

Butler, B. H., *see* 560.

200 **Butts, K. O.** 1971. Observations on the ecology of burrowing owls in western Oklahoma. A preliminary report. Proc. Okla. Acad. Sci. 51:66-74.—Study in Beaver and Texas counties indicated that burrowing owls are closely associated with *Cynomys ludovicianus* towns. Natural history, Ecology, Wildlife management. Sciuridae.

201 **Butts, K. O.** 1973. Life history and habitat requirements of burrowing owls in western Oklahoma. M.S. Thesis, Oklahoma State Univ., Stillwater. 188 pp.—Study in Beaver and Texas counties indicated that rodents are primary prey of burrowing owls in winter, and that the owls nest nearly exclusively in abandoned burrows of *Cynomys ludovicianus* and *Taxidea taxus.* Natural history, Ecology. Sciuridae, Mustelidae, Mammalia.

202 **Butts, K. O.** 1976. Burrowing owls wintering in the Oklahoma Panhandle. Auk 93:510-516.—Burrowing owls in Texas and Beaver counties associated closely with *Cynomys ludovicianus* towns, but most burrowing owls were gone during winter. Natural history, Ecology. Sciuridae.

203 **Butts, K. O., and J. C. Lewis.** 1982. The importance of prairie dog towns to burrowing owls in Oklahoma. Proc. Okla. Acad. Sci. 62:46-

52.—Burrowing owls (*Athene cunicularia*) scarce or absent in areas of central and western Oklahoma where black-tailed prairie dog (*Cynomys ludovicianus*) irradicated. Distribution, Natural history, Ecology, Wildlife management. Sciuridae.

Byard, M., *see* 287.

204 **Byers, A. M.** 1980. The Nature Conservancy. Endangered Species Tech. Bull. 5(6):10-11.—*Myotis grisescens* cave preserved in Oklahoma. Wildlife management. Vespertilionidae.

Cagle, F. G., *see* 164.

205 **Cahn, A. R.** 1932. Records and distribution of the fossil beaver, *Castoroides ohioensis*. J. Mammal. 13:229-241.—*Castoroides ohioensis* in Ottawa County. Paleontology. Castoridae.

206 **Caire, W.** 1986. Summer ecology of bats in southeastern Oklahoma. Nongame Wildlife Program Contract Rep., Oklahoma City. 63 pp.—Involved LeFlore, McCurtain, Pushmataha, and Choctaw counties. Distribution, Natural history, Ecology, Populations. Vespertilionidae.

207 **Caire, W., B. L. Cox, and B. Levescy.** 1981. Some normal blood values of *Myotis velifer* (Chiroptera: Vespertilionidae). J. Mammal. 62:436-439.—Blood values taken from *Myotis velifer* collected in hibernation. Physiology. Vespertilionidae.

208 **Caire, W., H. Haines, and T. M. McKenna.** 1982. Osmolality and ion concentration in urine of hibernating *Myotis velifer*. J. Mammal. 63: 688-690.—Cave myotis (*Myotis velifer*) studied from cave in Woodward County, Oklahoma. Physiology. Vespertilionidae.

209 **Caire, W., R. M. Hardisty, and K. E. Lacy.** 1986. Ecological notes on *Lasiurus cinereus* (Chiroptera: Vespertilionidae) in Oklahoma. Proc. Okla. Acad. Sci. 66:41-42.—Reports 16 captures of hoary bat, *Lasiurus cinereus,* in mist nets. Indicates capture times, sex, and reproductive condition. Distribution, Natural history, Ecology. Vespertilionidae.

210 **Caire, W., and L. Hornuff.** 1982. Wing morphology and flight behavior of the bat fly *Trichobius major* (Diptera: Streblidae). Southwest. Nat. 27:356-357.—Bat fly *Trichobius major* collected from *Myotis velifer* in Woodward County. Parasites. Vespertilionidae.

211 **Caire, W., L. Hornuff, and M. Ports.** 1981. Geographic variation in wing areas and femur lengths of the bat fly, *Trichobius major* (Diptera: Streblidae) in western Oklahoma. Southwest. Nat. 26:429-430.—Bat fly *Trichobius major* collected from *Myotis velifer* in Woodward and Washita counties. Parasites. Vespertilionidae.

212 **Caire, W., and M. A. Ports.** 1981. An adaptive method of predation by the great horned owl on Mexican free-tailed bats. Southwest. Nat. 26:69-70.—*Tadarida brasiliensis* captured in flight from behind by *Bubo virginianus.* Natural history. Molossidae.

213 **Caire, W., J. F. Smith, S. McGuire, and M. A. Royce.** 1984. Early foraging behavior of insectivorous bats in western Oklahoma. J. Mammal. 65:319-324.—Assessment of foraging activities of *Myotis velifer, Pipistrellus hesperus, Eptesicus fuscus, Plecotus townsendii,* and *Tadarida brasiliensis* in Woodward County. Behavior, Ecology. Vespertilionidae, Molossidae.

214 **Caire, W., and E. G. Zimmerman.** 1975. Chromosomal and morphological variation and circular overlap in the deer mouse, *Peromyscus maniculatus,* in Texas and Oklahoma. Syst. Zool. 24:89-95.—*Peromyscus maniculatus ozarkiarum, P. m. allescens,* and *P. m. luteus* in Ottawa, Adair, Pottawatomie, Pittsburg, Murray, Bryan, Dewey, Harmon, and Custer counties, Oklahoma, as well as Missouri, Arkansas, and Texas. Distinct chromosomal forms representing *P. m. ozarkiarum* and *P. m. pallescens* found. Gene flow restricted in eastern Oklahoma, west-central Texas, and eastern Texas. Systematics, Distribution, Genetics. Cricetidae.

Caire, W., *see* 661, 895.

215 **Cameron, G. N., E. D. Fleharty, and H. A. Watts.** 1979. Geographic variation in the energy content of cotton rats. J. Mammal. 60:817-820. —*Sigmodon hispidus* fat content varies geographically (Kansas, Oklahoma, Texas) and seasonally. Ecology, Physiology. Cricetidae.

216 **Cameron, G. N., and S. R. Spencer.** 1981. *Sigmodon hispidus.* Mamm. Species 158:1-9.—*Sigmodon hispidus* occurs statewide in Oklahoma. Summary of information in literature. Systematics, Paleontology, History, Distribution, Natural history, Behavior, Ecology, Populations, Genetics, Morphology, Physiology, Economics, Bibliographies. Cricetidae.

Cameron, G. N., *see* 1018.

217 **Caras, R. A.** 1967. North American mammals. Fur-bearing animals of the United States and Canada. Meredith Press, New York. 578 pp. —Systematics, Paleontology, History, Distribution, Natural history, Behavior, Ecology, Morphology, Techniques, Popular. Mammalia.

Carley, C., *see* 764.

218 **Carlile, F.** 1958. Deer investigations—northeast Oklahoma: Deer range densities. Okla. Dep. Wildl. Cons. Job 01, W-066-R-03. 15 pp. —Study of factors affecting *Odocoileus virginianus* populations in Cherokee, Delaware, and Adair counties. Natural history, Ecology, Populations, Techniques, Wildlife management. Canidae, Cervidae.

219 **Carlile, F.** 1958. Deer investigations—northeast Oklahoma: Vegetative study—refuges and outside range. Okla. Dep. Wildl. Cons. Job 05, W-066-R-03. 26 pp.—Study of *Odocoileus virginianus* habitats in Cherokee, Sequoyah, and Adair counties. Natural history, Ecology, Wildlife management. Cervidae.

220 **Carlile, F., and F. Lowry.** 1982. Elk monitoring. Okla. Dep. Wildl. Cons. Job 01, Study 01, W-80-R-21. 2 pp.—Elk (*Cervus elaphus*) sightings for northeastern Oklahoma. Distribution, Populations, Wildlife management. Cervidae.

Carlile, F., *see* 289, 1046, 1118.

221 **Carpenter, J. R.** 1937. Fluctuations in biotic communities, III. Aspection in a ravine sere in central Oklahoma. Ecology 18:80-92.—Characterizes plants and animals of ravines in central Oklahoma. Predominant mammals listed. Ecology. Mammalia.

222 **Carpenter, J. W., J. C. Freeney, and C. S. Patton.** 1972. Occurrence of *Demodex* Owen 1843 on a white-tailed deer from Oklahoma. J. Wildl. Dis. 8:112-114.—*Odocoileus virginianus* from Delaware County with infestation of *Demodex* mites. Parasites, Wildlife management. Cervidae.

223 **Carpenter, J. W., H. E. Jordan, and J. A. Morrison.** 1972. Meningeal worm (*Parelaphostrongylus tenuis*) infection in white-tailed deer in Oklahoma. J. Wildl. Dis. 8:381-383.—About half of *Odocoileus virginianus* examined from Pittsburg, Cherokee, Adair, and Delaware counties infected, and none from Osage or Woodward counties infected. Parasites, Wildlife management. Cervidae.

224 **Carpenter, J. W., H. E. Jordan, and B. C. Ward.** 1973. Neurologic disease in wapiti naturally infected with meningeal worms. J. Wildl. Dis. 9:148-153.—Natural infections of meningeal worm *Parelaphostrongylus tenuis* associated with neurologic disease in *Cervus canadensis* (=*elaphus*) from eastern Oklahoma. Worms not found in wapiti from Comanche County. Parasites. Cervidae.

225 **Carpenter, J. W., and R. P. Martin.** 1969. Capturing prairie dogs for transplanting. J. Wildl. Manage. 33:1024.—*Cynomys ludoviciana* (=*ludovicianus*). Pressurized water used to capture 29 prairie dogs in

Oklahoma. Technique simple, inexpensive, efficient and can probably be used with success in capturing other burrowing mammals. Techniques, Wildlife management. Sciuridae.

226 **Carter, D. C., W. D. Webster, J. K. Jones, Jr., C. Jones, and R. D. Suttkus.** 1985. *Dipodomys elator.* Mamm. Species 232:1-3.—Summary of published information on Texas kangaroo rat (*Dipodomys elator*). Systematics, History, Distribution, Natural history, Behavior, Ecology, Populations, Genetics, Morphology, Bibliographies. Heteromyidae.

Carter, D. C., *see* 308, 643, 644, 645.

227 **Cartledge, T. R.** 1970. The Tyler-Rose site and Late Prehistory in east-central Oklahoma. Okla. River Basin Surv. Archaeol. Site Rep. 19:1-182.—Strikingly high number (at least 56) of hispid cotton rats (*Sigmodon hispidus*) recovered. White-tailed deer (*Odocoileus virginianus*) and bison (*Bison bison*) found in small numbers. Paleontology. Cricetidae, Cervidae, Bovidae, Mammalia.

Castro, A. E., *see* 608.

228 **Chapman, J. A., and G. A. Feldhamer.** 1981. *Sylvilagus aquaticus.* Mamm. Species 151:1-4.—*Sylvilagus aquaticus aquaticus* in eastern half of Oklahoma. Summary of information in literature. Systematics, Paleontology, Distribution, Natural history, Behavior, Ecology, Populations, Parasites, Genetics, Morphology, Physiology. Leporidae.

229 **Chapman, J. A., and G. A. Feldhamer, eds.** 1982. Wild mammals of North America: Biology, management, and economics. Johns Hopkins Univ. Press, Baltimore. 1147 pp.—Detailed accounts of mammal species or groups of species with management implications. Emphasis on ecology and management, but life-history information also included. Systematics, History, Distribution, Natural history, Ecology, Economics, Wildlife management. Mammalia.

230 **Chapman, J. A., J. G. Hockman, and M. M. Ojeda C.** 1980. *Sylvilagus floridanus.* Mamm. Species 136:1-8.—Distribution of *Sylvilagus floridanus* includes Oklahoma. Summary of information in literature. Systematics, Paleontology, Distribution, Natural history, Behavior, Ecology, Populations, Parasites, Genetics, Morphology, Physiology, Economics, Bibliographies. Leporidae.

231 **Chapman, J. A., and G. R. Willner.** 1978. *Sylvilagus audubonii.* Mamm. Species 106:1-4.—Distribution of *Sylvilagus audubonii* includes western Oklahoma. Summary of information in literature. Systematics, Paleontology, Distribution, Natural history, Behavior,

Ecology, Populations, Parasites, Genetics, Morphology, Physiology, Economics, Bibliographies. Leporidae.

Chapman, J. A., *see* 1177.

Chappabitty, F. L., *see* 13.

232 **Chase, H. D.** 1939. Two species of mammals from Cimarron County new for Oklahoma. Proc. Okla. Acad. Sci. 19:69.—*Peromyscus truei truei* and *Erethizon epixanthum* (=*dorsatum*) *epixanthum* in Cimarron County. Distribution, Ecology. Cricetidae, Erethizontidae.

233 **Chase, H. D., and B. D. Barclay.** 1940. A species of *Neotoma* new for Oklahoma. Proc. Okla. Acad. Sci. 20:37.—*Neotoma mexicana fallax* in Cimarron County. Distribution. Cricetidae.

234 **Cheatham, J.** 1956. Southeastern Oklahoma game census: Vegetative study—refuges. Okla. Dep. Wildl. Cons. Job 06, W-061-R-01. 7 pp.—Study of *Odocoileus virginianus* utilization of game refuges in Pushmataha and Atoka counties. Natural history, Ecology, Populations, Wildlife management. Cervidae.

235 **Cheatham, J.** 1957. Southeastern Oklahoma game census: Vegetative study inside and outside refuges. Okla. Dep. Wildl. Cons. Job 06, W-061-R-02. 6 pp.—Study of *Odocoileus virginianus* habitats in Pushmataha and Atoka counties. Natural history, Ecology, Wildlife management. Cervidae.

Cheatum, E. P., *see* 965.

236 **Cheek, C. D., S. Bearden, W. Young, and C. B. Bousman.** 1981. Archeological investigations at the Graham site, 34At-90: 1975, 1977, and 1978. Report for Heritage Conservation and Recreation Service, U.S. Dep. Interior. Archeological Research Associates, Tulsa, Oklahoma. Research Rep. No. 23. 393 pp.—Bone faunal remains included several mammals. Paleontology. Leporidae, Cervidae, Bovidae.

237 **Chesemore, D. L.** 1971. Squirrel status study. Outdoor Okla. 27(7):15-17.—*Sciurus niger* and *S. carolinensis.* Account for the layman. Distribution, Techniques, Popular. Sciuridae.

238 **Chesemore, D. L.** 1972. In search of the squirrel. Outdoor Okla. 28(5):6-9.—Discussion of habits and habitats of *Sciurus niger* and *S. carolinensis.* Natural history, Behavior, Ecology, Techniques, Popular. Sciuridae.

239 **Chesemore, D. L.** 1975. Ecology of fox and gray squirrels (*Sciurus niger* and *Sciurus carolinensis*) in Oklahoma. Ph.D. Dissertation, Oklahoma State Univ., Stillwater. 348 pp.—Study of *Sciurus niger* and *S. carolinensis* in Creek, Okfuskee, and Okmulgee counties. Distribution, Natural history, Behavior, Ecology, Populations, Morphology, Physiology, Wildlife management. Sciuridae.

240 **Chesser, R. K., and M. L. Kennedy.** 1976. Predation on the free-tailed bat by the great horned owl. Bull. Okla. Ornithol. Soc. 9:1-3.—*Bubo virginianus* preyed on *Tadarida brasiliensis* in Greer County. Natural history, Ecology. Molossidae.

Chesser, R. K., *see* 86, 961, 1065.

Chichester, G., *see* 460.

241 **Choate, J. R., E. K. Boggess, and F. R. Henderson.** 1982. History and status of the black-footed ferret in Kansas. Trans. Kans. Acad. Sci. 85:121-132.—Information on endangered species (*Mustela nigripes*) in adjacent area. History, Distribution, Natural history, Populations. Mustelidae.

242 **Choate, J. R., C. J. Phillips, and H. H. Genoways.** 1966. Taxonomic status of the brush mouse, *Peromyscus boylii cansensis* Long, 1961. Trans. Kans. Acad. Sci. 69:306-313.—Comments on subspecific status of *Peromyscus boylii* in Oklahoma. Systematics, Distribution, Morphology. Cricetidae.

Choate, J. R., *see* 84, 341, 391, 392, 637, 642, 815, 910, 1065, 1208.

243 **Christensen, M. T.** 1974. Comparative aggressive behavior of wild and genetically sterile Norway rats, *Rattus norvegicus.* M.S. Thesis, Univ. Oklahoma, Norman. 40 pp.—Study includes *Rattus norvegicus* from Oklahoma County. Natural history, Behavior, Genetics, Economics. Muridae.

244 **Christensen, M. T.** 1977. Agonistic behavior between wild and genetically sterile Norway rats (*Rattus norvegicus*). Aggressive Behav. 3:17-31.—Genetically sterile male *Rattus norvegicus* more aggressive than wild rats from Oklahoma County. Natural history, Behavior, Genetics, Physiology. Muridae.

Christensen, M. T., *see* 697.

245 **Churcher, C. S.** 1959. The specific status of the New World red fox. J. Mammal. 40:513-520.—*Vulpes fulva* is a synonym for *V. vulpes.* The red fox shows clinal variation. Systematics, History, Distribution, Morphology. Canidae.

Ciskowski, C., *see* 461.

246 **Clark, C.** 1984. Archeological evidence for mountain lion (*Felis concolor*) at Lake Wister. Okla. Anthropol. Soc. Newsl. 32(6):4-5.—Mandible found at Wister Lake. Paleontology. Felidae.

247 **Clark, T. W.** 1965. Mammals of Boiling Springs State Park, Woodward County, Oklahoma. Proc. Okla. Acad. Sci. 46:36-38.—Lists 21 species of mammals. Distribution, Ecology. Mammalia.

248 **Clark, T. W.** 1965. Stomach contents of a whitetail deer fawn (*Odocoileus virginianus*). Proc. Okla. Acad. Sci. 46:220-221.—*Odocoileus virginianus* from Woodward County. Natural history, Ecology, Morphology. Cervidae.

249 **Clark, T. W.** 1967. Habitat selection by transplanted mule deer in northwest Oklahoma. Okla. Dep. Wildl. Cons. Jobs 01, 02, W-088-R-01. 49 pp.—Study of transplant populations of *Odocoileus hemionus* in Woodward, Harper, and Major counties. Distribution, Natural history, Behavior, Ecology, Morphology, Techniques, Wildlife management. Cervidae.

250 **Clark, T. W.** 1967. Habitat selection by transplanted mule deer in northwestern Oklahoma. M.S. Thesis, Oklahoma State Univ., Stillwater. 44 pp.—Study of *Odocoileus hemionus* in Cimarron, Harper, Woodward, and Major counties indicates lack of success for deer released east of Cimarron County. Distribution, Natural history, Behavior, Ecology, Populations, Wildlife management. Cervidae.

251 **Clark, T. W.** 1967. The life-form concept and mule deer habitat in Oklahoma. Proc. Okla. Acad. Sci. 48:23-27.—*Odocoileus hemionus hemionus* in Cimarron County. Presence of plant life-form elements used to identify mule deer habitat. Distribution, Ecology. Cervidae.

252 **Clark, T. W.** 1968. Plants used as food by mule deer in Oklahoma in relation to habitat. Southwest. Nat. 13:159-166.—*Odocoileus hemionus hemionus.* Distribution, Natural history, Ecology. Cervidae.

253 **Clark, T. W., and D. D. Skryja.** 1967. Further notes on the mammals of Black Mesa region, Oklahoma. Proc. Okla. Acad. Sci. 48:27-29.—Verified presence of eight previously unreported species in Cimarron County. Distribution, Ecology. Mammalia.

Clark, T. W., *see* 582.

254 **Clark, W. K.** 1953. The baculum in the taxonomy of *Peromyscus boylii* and *P. pectoralis.* J. Mammal. 34:189-192.—*Peromyscus boylii* and *P. pectoralis* may be identified with help of bacular morphology.

Some Oklahoma specimens of *P. boylii* used. Systematics, Morphology, Techniques. Cricetidae.

255 Clark, W. K. 1953. Gray shrew, *Notiosorex*, from eastern Oklahoma. J. Mammal. 34:117.—*Notiosorex crawfordi crawfordi* in Pushmataha County. Distribution, Natural history, Ecology, Morphology. Soricidae.

256 Cleveland, A. G. 1970. The current geographic distribution of the armadillo in the United States. Tex. J. Sci. 22:90-92.—*Dasypus novemcinctus* occurred throughout southeastern three-fifths of Oklahoma. History, Distribution, Ecology. Dasypodidae.

257 Cleveland, A. G. 1986. First record of *Baiomys taylori* north of the Red River. Southwest. Nat. 31:547.—Northernmost record of northern pygmy mouse. Distribution. Cricetidae.

258 Clymer, B. C., D. E. Howell, and J. A. Hair. 1970. Animal hosts of economically important ticks (Acarina) in east-central Oklahoma. Ann. Entomol. Soc. Am. 63:612-614.—Ticks of several species found on various bird and mammal species. Parasites. Mammalia.

259 Cockrum, E. L. 1951. A new pocket mouse (genus *Perognathus*) from Kansas. Univ. Kans. Publ. Mus. Nat. Hist. 5:203-206.—*Perognathus flavus bunkeri* distribution included western Oklahoma. Systematics, Distribution, Morphology. Heteromyidae.

260 Cockrum, E. L. 1952. The big free-tailed bat in Oklahoma. J. Mammal. 33:492.—*Tadarida molossa* (=*macrotis*) in Texas County. Distribution. Molossidae.

261 Cockrum, E. L. 1952. Mammals of Kansas. Univ. Kans. Publ. Mus. Nat. Hist. 7:1-303.—Covers area just to north of Oklahoma. Systematics, Paleontology, History, Distribution, Natural history, Behavior, Ecology, Parasites, Morphology, Techniques, Economics, Wildlife management, Bibliographies. Mammalia.

262 Cockrum, E. L. 1969. Migration in the guano bat, *Tadarida brasiliensis.* Univ. Kans. Mus. Nat. Hist. Misc. Publ. 51:303-336.—Oklahoma population of *Tadarida brasiliensis* is part of one of four separate North American breeding groups. Systematics, Distribution, Natural history, Behavior, Genetics, Techniques. Molossidae.

Cockrum, E. L., *see* 475.

263 Coggins, J. R., and J. S. McDaniel. 1975. Helminth population dynamics in the cotton rat, *Sigmodon hispidus.* Proc. Okla. Acad. Sci. 55:112-118.—Helminths occurred in 88% of *Sigmodon hispidus* studied. Cestodes and nematodes found. Seasonal variation in worm abundance, number of worm species, and host numbers. Males had heavier infestations. Parasites. Cricetidae.

Collier, G., *see* 292.

264 **Collins, S. L., and G. E. Uno.** 1983. The effect of early spring burning on vegetation in buffalo wallows. Bull. Torrey Bot. Club 110:474-481.—Analysis of vegetation in bison (*Bison bison*) wallows. Ecology. Bovidae.

Collins, S. L., *see* 885.

265 **Cones, E., and J. A. Allen.** 1877. Monographs of North American Rodentia. Rep. U.S. Geol. Surv. of the Territories 9:1-109.—Synopsis of known information concerning rodents and lagomorphs of North America, including some species from Indian Territory. Systematics, Distribution, Natural history. Leporidae, Sciuridae, Geomyidae, Castoridae, Cricetidae.

266 **Cook, H. J.** 1927. New geological and paleontological evidence bearing on the antiquity of mankind in America. Nat. Hist. 27:240-247.—Excavation of Tillman County site produced fragments of fossil mammoths, sloths, glyptodons, and camels. Paleontology. Mammalia.

267 **Cook, H. J.** 1927. New trails of early man in America. Sci. Am. 137:114-117.—Discussion of artifacts and fossil mammoths, mastodons, ground sloths, camels, and horses found at Tillman County site. Paleontology, Popular. Mammalia.

Cook, H. J., *see* 549, 550.

268 **Cope, E. D.** 1893. A new Pleistocene saber-tooth. Am. Nat. 27:896-897.—Type locality of *Dinobastis serus* in western Oklahoma, and brief mention of *Smilodon* species and *Elephas primigenius.* Systematics, Paleontology. Felidae, Elephantidae.

269 **Copelin, F. F.** 1959. Northwestern game census: Evaluation of the greater prairie chicken habitat management program, winter food plots. Okla. Dep. Wildl. Cons. Job 14, W-062-R-03. 3 pp.—Habitat management study includes mention of *Sylvilagus floridanus* and *Sigmodon hispidus.* Natural history, Ecology, Wildlife management. Leporidae, Cricetidae.

270 **Copelin, F. F.** 1960. Prairie chicken management research: Evaluation of the greater prairie chicken habitat management program winter food plots. Okla. Dep. Wildl. Cons. Job 14, W-062-R-04. 3 pp.—Habitat management study includes mention of *Sylvilagus* and

several rodent species. Natural history, Ecology, Wildlife management. Leporidae, Mammalia.

Copelin, F. F., *see* 1175.

271 **Cornely, J. E., and R. J. Baker.** 1986. *Neotoma mexicana.* Mamm. Species 262:1-7.—Summary of published information on Mexican woodrat, *Neotoma mexicana.* Systematics, Paleontology, History, Distribution, Natural history, Behavior, Ecology, Populations, Parasites, Genetics, Morphology, Bibliographies. Cricetidae.

Corner, R. G., *see* 964.

272 **Cothran, E. G.** 1982. Karyotypic, morphologic, and genic differentiation of hybridizing ground squirrels (*Spermophilus mexicanus* and *Spermophilus tridecemlineatus*). Ph.D. Dissertation, Univ. Oklahoma, Norman. 186 pp.—Study of hybridizing *Spermophilus mexicanus* and *S. tridecemlineatus* includes Cleveland County locality. Systematics, Distribution, Genetics, Morphology. Sciuridae.

273 **Cothran, E. G.** 1983. Morphologic relationships of the hybridizing ground squirrels *Spermophilus mexicanus* and *S. tridecemlineatus.* J. Mammal. 64:591-602.—Hybridization study includes sample of *Spermophilus tridecemlineatus* from Cleveland County. Systematics, Distribution, Morphology. Sciuridae.

274 **Cothran, E. G., and E. G. Zimmerman.** 1985. Electrophoretic analysis of the contact zone between *Geomys breviceps* and *Geomys bursarius.* J. Mammal. 66:489-497.—Electrophoretic analysis of contact zone in Oklahoma between pocket gophers that on basis of electrophoresis and karyology considered to be specifically distinct. Systematics, Distribution, Genetics. Geomyidae.

275 **Cowley, L. A.** 1974. Wildlife portrait: The cottontail. Outdoor Okla. 30(3):20.—*Sylvilagus floridanus.* Account for the layman. Distribution, Natural history, Behavior, Ecology, Morphology, Wildlife management, Popular. Leporidae.

276 **Cowley, L. A.** 1974. Wildlife portrait: The coyote. Outdoor Okla. 30(2):20.—*Canis latrans.* Account for the layman. Distribution, Natural history, Behavior, Ecology, Popular. Canidae.

277 **Cowley, L. A.** 1974. Wildlife portrait: The pronghorn. Outdoor Okla. 30(6):20.—*Antilocapra americana.* Account for the layman. Systematics, History, Distribution, Natural history, Behavior, Ecology, Morphology, Popular. Antilocapridae.

278 **Cowley, L. A.** 1976. Wildlife portrait: The badger. Outdoor Okla. 32(9):20.—*Taxidea taxus.* Account for the layman. Distribution, Natural history, Behavior, Ecology, Morphology, Popular. Mustelidae.

Cowley, L. A., *see* 338.

Cox, B. L., *see* 207.

Cozad, G. C., *see* 183.

279 **Crabb, E. D.** 1921. Some unexpected findings in the stomachs of predatory birds and mammals. Proc. Okla. Acad. Sci. 1:65-66.—Stomach contents of *Felis rufa* (=*rufus*) and *Mephitis mesomelas* (=*mephitis*) included no vertebrate food items. Natural history. Mustelidae, Felidae.

280 **Crabb, E. D.** 1922. A preliminary note on the number of tines in the antlers of the white-tail deer as correlated with agė. Proc. Okla. Acad. Sci. 2:12-13—Number of tines on *Odocoileus virginianus* antlers not correlated with age. Natural history, Morphology. Cervidae.

281 **Crabb, E. D.** 1924. The gross weight of Woodhouse's wolf. J. Mammal. 5:199-200.—*Canis latrans.* Morphology. Canidae.

282 **Crabb, E. D.** 1924. The occurrence of *Bison occidentalis* in Oklahoma. Proc. Okla. Acad. Sci. 4:124.—*Bison occidentalis* in Red Beds alluvium near Yukon, Canadian County. Paleontology, Distribution. Bovidae.

283 **Crabb, E. D.** 1924. The weight of an adult "coyote." Proc. Okla. Acad. Sci. 4:43.—*Canis frustror* (=*latrans*) from Comanche County. Morphology. Canidae.

284 **Cross, H.** 1917. Animal life in Oklahoma. Okla. Geol. Surv. Circ. 6:1-36.—Checklist and account for the layman of mammals and other vertebrates of Oklahoma. Natural history, Popular. Mammalia.

285 **Cross, H.** 1917. Animal life in Oklahoma. Okla. Geol. Surv. Bull. 27:183-213.—Checklist and account for the layman of mammals and other vertebrates of Oklahoma. Natural history, Popular. Mammalia.

286 **Currier, M. J. P.** 1983. *Felis concolor.* Mamm. Species 200:1-7.—Summary of published information on *Felis concolor,* including historical and recent distributions in Oklahoma. Systematics, Paleontology, History, Distribution, Natural history, Behavior, Ecology, Populations, Parasites, Genetics, Morphology, Economics, Bibliographies. Felidae.

287 **Curry, D., M. Byard, and R. Hatcher.** 1979. Small game population trends (inventory) Okla. Dep. Wildl. Cons. Job 01, Study 01, W-82-R-18. 13 pp.—Includes summary information on survey for squirrels, rabbits, raccoons, red foxes and bobcats. Distribution, Ecology, Populations. Leporidae, Sciuridae, Canidae, Procyonidae, Felidae.

Curry, D., *see* 562, 563.

288 **Curtis, C.** 1966. Deer season report. Okla. Dep. Wildl. Cons., Oklahoma City.—Distribution, Populations, Wildlife management. Cervidae.

289 **Curtis, C., F. C. Lowry, and F. Carlile.** 1971. Big game report 1971. Okla. Dep. Wildl. Cons. Pittman-Robertson Rep., Oklahoma City.—Distribution, Populations, Wildlife management. Cervidae, Mammalia.

290 **Cutter, W. L.** 1959. Notes on some mammals from northern Texas. Southwest. Nat. 4:30-34.—Mentions observing swift foxes (*Vulpes velox*) in Texas County, Oklahoma. Distribution. Canidae.

291 **Dalquest, W. W.** 1977. Mammals of the Holloman local fauna, Pleistocene of Oklahoma. Southwest. Nat. 22:255-268.—Early Irvingtonian records of *Equus, Mammuthus, Stegomastodon, Megalonyx, Glossotherium, Glyptotherium, Cynomys, Felis, Platygonus, Titanotylopus, Ovibos* or *Symbos, Camelops, Hemiauchenia, Odocoileus,* and *Tapirus.* Paleontology, Distribution. Mammalia.

292 **Dalquest, W. W., and G. Collier.** 1964. Notes on *Dipodomys elator,* a rare kangaroo rat. Southwest. Nat. 9:146-150.—Comanche County and five Texas counties. *Dipodomys elator* closely related to *D. ornatus.* Also mentions *D. ordii, D. phillipsi, D. spectabilis,* and *D. merriami.* Systematics, Distribution, Ecology, Morphology. Heteromyidae.

293 **Dalquest, W. W., and N. V. Horner.** 1984. Mammals of north central Texas. Midwestern State Univ. Press, Wichita Falls, Texas. 261 pp.—Mammals of area adjacent to Oklahoma. Systematics, History, Distribution, Natural history, Morphology. Mammalia.

Dalquest, W. W., *see* 476, 1024.

Danielson, R. N., *see* 1161.

Davis, E. M., *see* 1124.

294 **Davis, W. B.** 1940. Distribution and variation of pocket gophers (genus *Geomys*) in the southwestern United States. Tex. Agric. Exp.

Stn. Bull. 590:1-38.—*Geomys breviceps dutcheri,* new subspecies, occurs in eastern Oklahoma, with type locality in Muskogee County. *Geomys lutescens major* occurs in western Oklahoma. Systematics, Distribution, Morphology. Geomyidae.

295 **Davis, W. B.** 1942. The moles (genus *Scalopus*) of Texas. Am. Midl. Nat. 27:380-386.—Systematics of *Scalopus aquaticus* of Texas, including areas along Oklahoma border. Systematics, Distribution. Talpidae.

296 **Davis, W. B.** 1942. The systematic status of four kangaroo rats. J. Mammal. 23:328-333.—*Dipodomys oklahomae* referred to *Dipodomys ordii.* Systematics. Heteromyidae.

297 **Davis, W. B.** 1974. The mammals of Texas. Tex. Parks Wildl. Dep. Bull. 41:1-294.—Includes distribution maps of all Texas mammals; covers area adjacent to Oklahoma. Distribution, Natural history, Behavior, Ecology, Wildlife management. Mammalia.

Davis, W. B., *see* 311.

298 **Davis, W. H.** 1959. Taxonomy of the eastern pipistrel. J. Mammal. 40:521-531.—*Pipistrellus s. subflavus* distribution includes Oklahoma excluding Panhandle. Systematics, Distribution. Vespertilionidae.

Davis, W. H., *see* 104.

299 **Davy, R. H.** 1956. The black-tailed prairie dog (*Cynomys ludovicianus*) as a member of the biotic community. M.S. Thesis, Oklahoma A and M Coll., Stillwater. 101 pp.—Ecology. Sciuridae.

300 **Day, S. D.** 1978. A survey of furbearers and trappers in Oklahoma. M.S. Thesis, Oklahoma State Univ., Stillwater. 104 pp.—Statewide analysis of furbearer populations as indicated by trapping reports and fur sales. Distribution, Populations, Economics, Wildlife management. Mammalia.

301 **DeArment, R.** 1951. Habitat improvement evaluation: Determination of plant survival in habitat plots. Okla. Dep. Wildl. Cons. Job 01, W-044-R-01. 15 pp.—Habitat management study includes discussion of *Lepus californicus, Sylvilagus, Geomys, Dipodomys, Neotoma,* and *Sigmodon hispidus.* Natural history, Ecology, Economics, Wildlife management. Leporidae, Geomyidae, Heteromyidae, Cricetidae.

302 **DeArment, R.** 1951. Habitat improvement evaluation: Determination of utilization of completed plots by bobwhite quail and other wildlife. Okla. Dep. Wildl. Cons. Job 03, W-044-R-01. 12 pp.—Habitat management study includes discussion of *Sylvilagus, Sciurus,*

Geomys, and *Mephitis mephitis.* Natural history, Ecology, Techniques, Economics, Wildlife management. Leporidae, Geomyidae, Sciuridae, Mustelidae, Mammalia.

303 **Decker, C. E.** 1924. A preliminary list giving the distribution of Proboscidea in Oklahoma. Proc. Okla. Acad. Sci. 4:123-124.—Mastodons and *Elephas columbi* found at 22 localities. Paleontology, Distribution. Mastodontidae, Elephantidae.

304 **Desha, P. G.** 1966. Observations on the burrow utilization of the thirteen-lined ground squirrel. Southwest. Nat. 11:408-410.—*Citellus* (=*Spermophilus*) *tridecemlineatus* burrows evaluated near Willis. Squirrel use of burrows studied. Natural history, Behavior. Sciuridae.

305 **Dickson, K. L., and D. Vance.** 1981. Revegetating surface mined lands for wildlife in Texas and Oklahoma. U.S. Fish Wildl. Serv. FWS/OBS-81/25. 121 pp.—Discussion of revegetation objectives and methods for areas including eastern Oklahoma. Ecology, Wildlife management. Mammalia.

306 **Dillehay, T. D.** 1974. Late Quaternary bison population changes on the Southern Plains. Plains Anthropol. 19:180-196.—Discussion of changing distributions of *Bison* species during Quaternary. Paleontology, Distribution, Bibliographies. Bovidae.

307 **DiSilvestro, R., R. Umber, and N. Keyes.** 1980. The Oklahoma whitetail. Outdoor Okla. 36(9):3-11.—Discussion of habits, habitats and populations of *Odocoileus virginianus* in Oklahoma. History, Natural history, Behavior, Ecology, Populations, Techniques, Wildlife management, Popular. Cervidae.

308 **Dolan, P. G., and D. C. Carter.** 1977. *Glaucomys volans.* Mamm. Species 78:1-6.—Distribution of *Glaucomys volans* includes eastern Oklahoma. Summary of information in literature. Systematics, Paleontology, History, Distribution, Natural history, Behavior, Ecology, Populations, Parasites, Genetics, Morphology, Physiology, Bibliographies. Sciuridae.

309 **Dolgos, A. G., and G. A. Earls.** 1973. Observations on a nutria (*Myocaster coypus*) population in southeastern Oklahoma. Proc. Okla. Acad. Sci. 53:109-110.—*Myocaster* (=*Myocastor*) *coypus* in McCurtain County. Distribution, Natural history, Ecology, Populations, Morphology. Myocastoridae.

310 **Dort, W., Jr., and J. K. Jones, Jr.** 1970. Pleistocene and Recent environments of the central Great Plains. Spec. Publ. Dep. Geol. Univ. Kans. 3:1-433.—Contents include: Quaternary mammalian sequence

in the central Great Plains (C. B. Schultz and L. D. Martin); Influence of late-glacial and post-glacial events on the distribution of Recent mammals on the northern Great Plains (R. S. Hoffmann and J. K. Jones, Jr.); and Pleistocene mammalian local faunas from the Great Plains and Central Lowland provinces of the United States (C. W. Hibbard). Systematics, Paleontology, Distribution, Ecology. Mammalia.

311 **Doutt, J. K., A. B. Howell, and W. B. Davis.** 1945. The mammal collections of North America. J. Mammal. 26:231-272.—Systematics, Distribution. Mammalia.

Doutt, J. K., *see* 16.

Doyle, R., *see* 681, 607.

312 **Drabek, C. M.** 1977. Some population aspects of cricetid rodents in a central Oklahoma tall grass prairie. Proc. Okla. Acad. Sci. 57:25-28. —Live-trap study in which 22 mammal species (six cricetids) captured. Distribution, Ecology, Populations. Cricetidae, Mammalia.

313 **Drass, R. R.** 1979. Roulston-Rogers; a stratified plains woodland and Late Archaic site in the Cross Timbers. Bull. Okla. Anthropol. Soc. 28:1-135.—Bison (*Bison bison*), white-tailed deer (*Odocoileus virginianus*), and fox (probably *Vulpes vulpes*) recovered from site in Seminole County. Paleontology. Canidae, Cervidae, Bovidae.

Drass, R. R., *see* 178.

314 **Drescher, A. B.** 1939. A new Pliocene badger from Mexico. Bull. South. Calif. Acad. Sci. 38:57-62.—Fossil badger tooth from Texas County site tentatively allocated to *Taxidea mexicana.* Systematics, Paleontology, Morphology. Mustelidae.

315 **Duck, L. G., and J. B. Fletcher.** 1944. A survey of the game and furbearing animals of Oklahoma. Okla. Game Fish Comm. Pittman-Robertson Ser. II, State Bull. 3:1-144.—Evaluation of abundances and distributions of game mammals of Oklahoma. History, Distribution, Natural history, Techniques, Economics, Wildlife management. Mammalia.

316 **Due, L. A.** 1967. Fort Gibson game management area development: Herbaceous seedings. Okla. Dep. Wildl. Cons. Job 13, W-054-D-12. 3 pp.—Study of Muskogee County habitats with respect to *Odocoileus virginianus.* Natural history, Ecology, Techniques, Wildlife management. Cervidae, Mammalia.

317 **Duffield, L. F.** 1969. The vertebrate faunal remains from the School Land I and School Land II sites, Delaware County, Oklahoma. Bull. Okla. Anthropol. Soc. 18:47-65.—Sites in northeastern Oklahoma yielded remains of 10 mammal species, including domestic pigs and dogs. White-tailed deer (*Odocoileus virginianus*) found most frequently. Other large mammals include bison (*Bison bison*) and elk (*Cervus elaphus*). Paleontology. Cervidae, Bovidae, Mammalia.

318 **Duffield, L. F., and E. E. Pillaert.** 1967. Modern bison appeared on the Southern Plains 750 A.D.?? Okla. Anthropol. Soc. Newsl. 15(1): 3-8.—Discussion of prehistoric distribution of *Bison bison.* Paleontology. Bovidae.

319 **Dunbar, M. R.** 1973. Seasonal changes in testis morphology and spermatogenesis in adult and young-of-the-year coyotes (*Canis latrans*). M.S. Thesis, Oklahoma State Univ., Stillwater. 24 pp.—*Canis latrans* in Oklahoma has peak breeding season in February. Natural history, Morphology, Physiology, Techniques. Canidae.

320 **Dunbar, M. R.** 1976. Breeding season of white-tailed deer in eastern Oklahoma. Proc. Okla. Acad. Sci. 56:24-25.—*Odocoileus virginianus.* Natural history. Cervidae.

Dunbar, M. R., *see* 776.

Dunn, J. C., *see* 685.

Earls, G. A., *see* 309.

321 **Edwards, R. M.** 1977. Effects of ambient water vapor pressure and temperature on evaporative water loss in *Peromyscus maniculatus* and *Mus musculus.* M.S. Thesis, Univ. Oklahoma, Norman. 34 pp.—Evaporative water loss greater for *Peromyscus maniculatus* than for *Mus musculus.* Physiology. Cricetidae, Muridae.

322 **Edwards, R. M., and H. Haines.** 1978. Effects of ambient water vapor pressure and temperature on evaporative water loss in *Peromyscus maniculatus* and *Mus musculus.* J. Comp. Physiol. 128:177-184.—Used deer mice (*Peromyscus maniculatus*) captured near Norman, Oklahoma. Physiology. Cricetidae, Muridae.

Edwards, W. C., *see* 682.

323 **Egoscue, H. J.** 1979. *Vulpes velox.* Mamm. Species 122:1-5.—Distribution of *Vulpes velox* includes western Oklahoma. Summary of information in literature. Systematics, Paleontology, History, Distribution, Natural history, Behavior, Ecology, Parasites, Genetics, Morphology, Economics, Bibliographies. Canidae.

324 **Elder, W. H., and W. J. Gunier.** 1978. Sex ratios and seasonal movements of gray bats, *Myotis grisescens,* in southwestern Missouri and adjacent states. Am. Midl. Nat. 99:463-473.—*Myotis grisescens* in Adair, Muskogee, Cherokee, Ottawa, Delaware, and Craig counties, Oklahoma, in addition to Missouri, Kansas, Arkansas, and Illinois. Distribution, Natural history, Ecology, Populations. Vespertilionidae.

325 **Elliot, D. G.** 1899. Description of apparently new species and subspecies of mammals from the Indian Territory. Field Mus. Nat. Hist. Publ. Zool. Ser. 1:285-288.—Type localities of *Lepus temalemonus* (=*Sylvilagus aquaticus*) and *Blarina brevicauda hulophaga* (=*hylophaga*) in Murray County. Systematics, Distribution, Morphology. Soricidae, Leporidae.

326 **Elliot, D. G.** 1899. Description of apparently new species and subspecies of mammals from Oklahoma Territory. Field Mus. Nat. Hist. Publ. Zool. Ser. 1:279-282.—Type locality of *Neotoma macropus* (=*micropus*) *surberi* in Woods County, of *Scalops* (=*Scalopus*) *machrinus* (=*aquaticus*) *intermedius* in Woods County, and of *Reithrodontomys chrysotis* (=*fulvescens*) in Murray County. Systematics. Talpidae, Cricetidae.

327 **Elliot, D. G.** 1899. List of mammals obtained by Thaddeus Surber, collector for the Museum, chiefly in Oklahoma and Indian territories. Field Mus. Nat. Hist. Publ. Zool. Ser. 1:291-303.—Accounts of 25 species, and localities in Oklahoma and Indian territories. Systematics, Distribution, Natural history. Mammalia.

328 **Elliot, D. G.** 1901. A synopsis of the mammals of North America and the adjacent seas. Field Mus. Nat. Hist. Publ. Zool. Ser. 2:1-471.—Accounts of all species of mammals north of Mexico. Type locality given for each taxon. Systematics, Distribution, Morphology. Mammalia.

329 **Elliot, D. G.** 1904. Descriptions of twenty-seven apparently new species and subspecies of mammals, all but six collected by Edmund Heller. Field Mus. Nat. Hist. Publ. Zool. Ser. 3:239-261.—Type locality of *Perognathus* (=*Chaetodipus*) *hispidus maximus* (=*spilotis*) in Cleveland County. Systematics, Distribution. Heteromyidae.

330 **Elliot, D. G.** 1905. A check list of mammals of the North American continent, the West Indies and the neighboring seas. Field Mus. Nat. Hist. Publ. Zool. Ser. 6:1-761.—Accounts for all species in North and Middle America. Systematics, Distribution, Morphology. Mammalia.

331 **Elliot, D. G.** 1905. Descriptions of apparently new mammals of the genera *Ovibos, Cynomys,* and *Mustela.* Proc. Biol. Soc. Wash. 18:135-140.—Type locality of *Cynomys pyrrotrichus* (=*ludovicianus*) is White Horse Spring, Oklahoma Territory. Systematics, Distribution, Morphology. Sciuridae.

332 **Elliot, D. G.** 1907. A catalogue of the collection of mammals in the Field Columbian Museum. Field Columbian Mus. Publ. Zool. Ser. 8:1-694.—Lists specimens from Oklahoma, including type specimen of *Blarina hylophaga.* Systematics, Distribution. Soricidae, Mammalia.

Elliot, D. G., *see* 927.

Ellis, L. L., *see* 1164.

333 **Ellis, L. L., Jr.** 1952. The ectoparasites of certain mammals in Oklahoma, their ecology and disease relationships. Ph.D. Dissertation, Univ. Oklahoma, Norman. 94 pp.—Study of fleas, ticks, lice and mites of mammals of Wichita Mountains, Comanche County, revealed none carrying plague. Ecology, Populations, Parasites. Mammalia.

334 **Ellis, L. L., Jr.** 1952. A tentative key to the mammalian ectoparasites of the Wichita Mountains Wildlife Refuge. Proc. Okla. Acad. Sci. 33:111-115.—Key to 23 ectoparasite species from Wichita Mountains Wildlife Refuge, Comanche County. Total of 31 parasite species discussed. Parasites. Mammalia.

335 **Ellis, L. L., Jr.** 1955. A survey of the ectoparasites of certain mammals in Oklahoma. Ecology 36:12-18.—Study includes 31 species of ectoparasites and 29 species of mammals in Comanche County. Relatively low ectoparasite infestations noted. Ectoparasite list included. Ecology, Parasites, Techniques. Mammalia.

336 **Ellis, R. J.** 1958. Food habits and control of coyotes in northcentral Oklahoma. M.S. Thesis, Oklahoma State Univ., Stillwater. 46 pp.—*Canis latrans* in Noble and Payne counties ate mostly species of *Lepus* and *Sylvilagus,* as well as *Sigmodon hispidus.* Natural history, Wildlife management. Leporidae, Cricetidae, Canidae, Mammalia.

337 **Ellis, R. J.** 1959. Food habits and control of coyotes in northcentral Oklahoma. Okla. State Univ. Publ. Arts Sci. Stud. Biol. Ser. 8:1-31.—*Canis latrans* in Payne and Noble counties ate mostly species of *Sylvilagus* and *Lepus,* as well as *Sigmodon hispidus.* Natural history, Wildlife management. Leporidae, Cricetidae, Canidae, Mammalia.

338 **Ellis, R. J., and L. A. Cowley.** 1968. Wolf hunting is big business as practiced in Oklahoma. Hunter's Horn (Sand Springs, Oklahoma).—Economics, Wildlife management, Popular. Canidae.

339 **Ellis, R. J., and S. D. Schemnitz.** 1957. Some foods used by coyotes and bobcats in Cimarron County, Oklahoma, 1954 through 1956. Proc. Okla. Acad. Sci. 38:180-185.—Predator stomachs examined in Comanche County. Rabbits appeared to be staple food for *Canis latrans* and *Lynx* (=*Felis*) *rufus.* Bobcat diet included kangaroo rats, woodrats, grasshopper mice, pocket mice, birds, and grass. Coyotes ate cattle, kangaroo rats, lizards, insects, and grass. Ecology, Economics. Canidae, Felidae, Mammalia.

Engbretson, G. A., *see* 653.

340 **Engstrom, M. D.** 1978. Systematics of the northern grasshopper mouse (*Onychomys leucogaster*) on the central Great Plains. M.S. Thesis, Fort Hays State Univ., Hays, Kansas. 35 pp.—Distribution of *Onychomys leucogaster breviauritus* in western Oklahoma, and *O. l. articeps* in Oklahoma Panhandle. Systematics, Distribution, Natural history, Morphology. Cricetidae.

341 **Engstrom, M. D., and J. R. Choate.** 1979. Systematics of the northern grasshopper mouse (*Onychomys leucogaster*) on the central Great Plains. J. Mammal. 60:723-739.—*Onychomys leucogaster breviauritus* synonymous with *O. l. articeps.* Systematics, Distribution, Natural history, Morphology. Cricetidae.

342 **Erskine, D. J.** 1981. Critical thermal maxima and thermal tolerance in small mammals. Ph.D. Dissertation, Univ. Oklahoma, Norman. 48 pp.—Study includes *Peromyscus leucopus* and *Dipodomys ordii* from Cleveland County. Thermal tolerance was high for *Dipodomys* and medium for *Peromyscus.* Ecology, Physiology. Heteromyidae, Cricetidae.

343 **Erskine, D. J., and V. H. Hutchison.** 1982. Critical thermal maxima in small mammals. J. Mammal. 63:267-273.—Study includes *Peromyscus leucopus* and *Dipodomys ordii* from Cleveland County. Critical thermal maximum of *D. ordii* significantly higher than that for other species. Physiology. Heteromyidae, Cricetidae.

344 **Evans, J.** 1970. About nutria and their control. U.S. Fish Wildl. Serv. Resour. Publ. 86:1-65.—Populations of *Myocastor coypus* established in southern Oklahoma. Distribution, Natural history, Behavior, Ecology, Morphology, Techniques, Economics, Popular. Myocastoridae.

345 **Eve, H.** 1975. The enemy within: Research at Darlington will focus on parasites in wildlife. Outdoor Okla. 31(3):12-13.—Account for layman. Concerns parasites, primarily of white-tailed deer (*Odocoileus virginianus*) and elk (*Cervus elaphus*). Parasites, Popular. Cervidae.

346 **Eve, H.** 1975. A study of two Oklahoma deer herds. Parasite-based research shows white-tail overpopulation on Cookson and Pushmataha GMA's. Outdoor Okla. 31(8):6-9.—Account for the layman based on population study of *Odocoileus virginianus.* Parasites, Wildlife management, Popular. Cervidae.

347 **Eve, H.** 1978. A future for furbearers. Outdoor Okla. 34(11):2-6.—Discussion of populations and regulations of *Urocyon cinereoargenteus, Procyon lotor,* and *Felis rufus.* Populations, Wildlife management. Canidae, Procyonidae, Felidae.

348 **Eve, H.** 1978. Oklahoma's furbearer study. Outdoor Okla. 34(3):5-7.—Furbearer distributions and abundances evaluated by scent-post surveys, pond surveys, and fur-dealer reports. Populations, Techniques, Wildlife management, Popular. Procyonidae, Felidae.

349 **Eve, J. H.** 1978. Big game disease research. Okla. Dep. Wildl. Cons. Job 02, Study 03, W-80-R-17. 18 pp.—Baseline data on diseases and parasites of white-tailed deer (*Odocoileus virginianus*). Natural history, Parasites, Wildlife management. Cervidae.

350 **Eve, J. H.** 1981. Management implications of disease. Diseases and parasites of white-tailed deer (W. R. Davidson, ed.). Tall Timbers Research Station Misc. Publ., Tallahassee, Florida. 458 pp.—Uses some Oklahoma data to evaluate disease and management practices for *Odocoileus virginianus.* Ecology, Parasites, Wildlife management. Cervidae.

Eve, J. H., *see* 682.

351 **Ewing, L. L., P. M. Green, and A. M. Stebler.** 1965. Metabolic and biochemical changes in testis [testes] of cotton rat[s] (*Sigmodon hispidus*) during breeding cycle. Proc. Soc. Exp. Biol. Med. 118:911-913.—Hispid cotton rats (*Sigmodon hispidus*) collected monthly (total 91) from natural habitat. Variety of morphological and physiological measures taken of testes and accessory organs. Morphology, Physiology. Cricetidae.

352 **Farr, C. H.** 1952. Investigation of birth productivity in *Myotis.* Proc. Okla. Acad. Sci. 33:29-30.—*Myotis velifer* in Woodward and Major counties. Natural history. Vespertilionidae.

Feldhamer, G. A., *see* 228, 229, 1177.

353 **Fenton, M. B., and R. M. R. Barclay.** 1980. *Myotis lucifugus.* Mamm. Species 142:1-8.—*Myotis l. lucifugus* in northeastern two-thirds of Oklahoma. Summary of information in literature. Systematics, Paleontology, Distribution, Natural history, Behavior, Ecology, Populations, Parasites, Genetics, Morphology, Physiology. Vespertilionidae.

354 **Fenwood, J. D., D. F. Urbston, and R. F. Harlow.** 1984. Determining deer habitat capability in Ouachita National Forest pine stands. Proc. Annu. Conf. Southeast. Assoc. Fish Wildl. Agencies 38:13-22.—Timber stand age and basal area of shortleaf pine (*Pinus echinata*) related to forage yields for white-tailed deer (*Odocoileus virginianus*) in Arkansas and Oklahoma. Ecology, Wildlife management. Cervidae.

355 **Figgins, J. D.** 1927. The antiquity of man in America. Nat. Hist. 27:229-239.—Tillman County site produced specimens of *Mylodon, Equus, Trilophodon,* and *Elephas* in association with artifacts. Paleontology. Mylodontidae, Mammalia.

356 **Findley, J. S., A. H. Harris, D. E. Wilson, and C. Jones.** 1975. Mammals of New Mexico. Univ. New Mexico Press, Albuquerque. 360 pp. —Records of mammals along western Oklahoma border. Includes keys. Distribution, Natural history. Mammalia.

357 **Findley, J. S., and C. Jones.** 1964. Seasonal distribution of the hoary bat. J. Mammal. 45:461-470.—*Lasiurus cinereus* known to occur in Oklahoma during April, August, and September. Systematics, Distribution, Natural history. Vespertilionidae.

358 **Findley, J. S., and G. L. Traut.** 1970. Geographic variation in *Pipistrellus hesperus.* J. Mammal. 51:741-765.—Includes specimens from Comanche and Kiowa counties. Systematics, Distribution, Morphology. Vespertilionidae.

Findley, J. S., *see* 16.

359 **Finley, R. B., Jr.** 1953. A new subspecies of wood rat (*Neotoma mexicana*) from Colorado. Univ. Kans. Publ. Mus. Nat. Hist. 5:527-534.—New subspecies of *Neotoma mexicana.* Range does not include Oklahoma, but Cimarron County specimen used for comparison. Systematics, Distribution, Morphology. Cricetidae.

360 **Finley, R. B., Jr.** 1958. The wood rats of Colorado: Distribution and ecology. Univ. Kans. Publ. Mus. Nat. Hist. 10:213-552.—Some specimens from Oklahoma Panhandle of the southern plains woodrat (*Neotoma micropus*) used for comparison. Notes that Regnier,

Colorado and Regnier, Oklahoma are different but geographically close localities. Systematics, Distribution, Natural History, Ecology, Morphology. Cricetidae.

361 **Fitch, H. S., P. Goodrum, and C. Newman.** 1952. The armadillo in the southeastern United States. J. Mammal. 33:21-37.—*Dasypus novemcinctus* extending its range in United States. Distribution, Natural history, Ecology, Populations, Economics. Dasypodidae.

362 **Fitch, J. H., and K. A. Shump, Jr.** 1979. *Myotis keenii.* Mamm. Species 121:1-3.—Distribution of *Myotis keenii* includes northeastern Oklahoma. Summary of information in literature. Systematics, Distribution, Natural history, Behavior, Ecology, Parasites, Morphology, Physiology, Bibliographies. Vespertilionidae.

363 **Fitch, J. H., K. A. Shump, Jr., and A. U. Shump.** 1981. *Myotis velifer.* Mamm. Species 149:1-5.—*Myotis velifer grandis* in western two-fifths of Oklahoma, not including Panhandle. Summary of information in literature. Systematics, Paleontology, Distribution, Natural history, Behavior, Ecology, Populations, Parasites, Genetics, Morphology, Physiology, Bibliographies. Vespertilionidae.

Fitzgerald, J. P., *see* 1058, 1059.

Fleharty, E. D., *see* 215.

Fletcher, J., *see* 170, 465, 315, 688.

364 **Fletcher, J. R.** 1959. Cookson Hills Refuge development: Food plot usage. Okla. Dep. Wildl. Cons. Job 19/B, W-046-D-05. 6 pp.—Study of Cherokee County habitat for *Odocoileus virginianus.* Natural history, Ecology, Populations, Wildlife management. Cervidae.

365 **Flowers, T. L.** 1985. Recent breeding of the mountain plover in Cimarron County, Oklahoma. Bull. Okla. Ornithol. Soc. 18:10-12.—Mountain plover (*Charadrius montanus*) nests found in vicinity of active black-tailed prairie dog (*Cynomys ludovicianus*) colony. Ecology. Sciuridae.

366 **Flynn, P.** 1984. An analysis of the 1973 test excavations at the Zimms site (34RM72). Pages 215-290 *in* Archaeology of the mixed grass prairie. Phase 1: Quartermaster Creek. Okla. Archeol. Surv., Okla. Archeol. Resource Surv. Rep. No. 20. 313 pp.—Faunal remains included at least five mammal species. Paleontology. Leporidae, Sciuridae, Mustelidae, Cervidae, Bovidae.

Flynn, P., *see* 1131.

Foreman, D., *see* 80.

367 **Foster, B.** 1961. Prehistoric elephant skeleton found north of Lawton. Okla. Anthropol. Soc. Newsl. 9(4):3-5.—Skeleton of mammoth or, less probably, mastodon found in Comanche County. Paleontology. Mastodontidae, Elephantidae.

368 **Frank, W. J.** 1950. Rodent populations and their reactions to grazing intensities on sand sagebrush grasslands in the southern Great Plains region. Ph.D. Dissertation, Oklahoma State Univ., Stillwater. 204 pp.—Study of rodent and rabbit populations in Harper and Woodward counties. Ecology, Populations, Techniques. Mammalia.

Free, P., *see* 765.

369 **Freeman, R. C.** 1976. Coyote x dog hybridization and red wolf influence in the wild *Canis* of Oklahoma. M.S. Thesis, Oklahoma State Univ., Stillwater. 62 pp.—Study of wild *Canis* indicates primarily *C. latrans,* with some introgression from *C. rufus* in the southeast and *C. familiaris* statewide. Populations, Morphology, Economics. Canidae.

370 **Freeman, R. C.** 1977. A wolf by any other name. Outdoor Okla. 33(6):17-19.—*Canis lupus nubilus, C. rufus,* and *C. latrans.* Account for the layman on wild canid populations in Oklahoma. Coyotes, coyote-dog hybrids and coyote-red wolf hybrids. Distribution, Natural history, Morphology, Popular. Canidae.

371 **Freeman, R. C., and J. H. Shaw.** 1979. Hybridization in *Canis* (Canidae) in Oklahoma. Southwest. Nat. 24:485-499.—Morphological evidence for genetic introgression of *Canis familiaris* and *C. rufus* into *C. latrans* populations. Populations, Genetics. Canidae.

Freeney, J. C., *see* 222.

372 **French, N. R.** 1971. Small mammal studies in the U.S.—IBP grassland biome. Ann. Zool. Fenn. 8:48-53.—Includes data from Osage site, Osage County. Ecology, Populations. Mammalia.

373 **French, N. R., and W. E. Grant.** 1974. Summary report of small mammal project grid live-trapping data. U.S. IBP (Int. Biol. Program) Grassland Biome Tech. Rep. 258:1-27. Colorado State Univ., Fort Collins.—Summary of trapping information for small mammals at Osage County site. Ecology, Populations, Techniques. Soricidae, Sciuridae, Heteromyidae, Cricetidae, Muridae.

French, N. R., *see* 450.

374 **Fritzell, E. K., and K. J. Haroldson.** 1982. *Urocyon cinereoargenteus.* Mamm. Species 189:1-8.—Distribution of *Urocyon cinereoargenteus* includes most of Oklahoma. Summary of information in literature. Systematics, Paleontology, Distribution, Natural history, Behavior, Ecology, Populations, Parasites, Genetics, Morphology, Physiology, Bibliographies. Canidae.

375 **Frye, R. G.** 1973. Wildlife-oriented recreational opportunities on public school land in Payne County, Oklahoma. M.S. Thesis, Oklahoma State Univ., Stillwater. 110 pp.—Payne County study includes abundance estimates of some mammals. Ecology, Techniques, Wildlife management. Mammalia.

376 **Fujita, M. S., and T. H. Kunz.** 1984. *Pipistrellus subflavus.* Mamm. Species 228:1-6.—Summary of information in literature. *Pipistrellus subflavus* occurs statewide except in Panhandle. Systematics, Paleontology, History, Distribution, Natural history, Behavior, Ecology, Populations, Parasites, Genetics, Morphology, Economics, Bibliographies. Vespertilionidae.

Furman, D. P., *see* 896.

377 **Galm, J. R.** 1981. Prehistoric cultural adaptations in the Wister Valley, east-central Oklahoma. Ph.D. Dissertation, Washington State Univ., Pullman. 274 pp.—Information from three study sites used to define seven cultural phases from 300 B.C. to 1907 A.D. Dominant foods included white-tailed deer (*Odocoileus virginianus*) and small game. Fifteen mammal species identified. Paleontology. Cervidae, Mammalia.

Galm, J. R., *see* 166.

378 **Gardner, A. L.** 1973. The systematics of the genus *Didelphis* (Marsupialia: Didelphidae) in North and Middle America. Spec. Publ. Mus. Tex. Tech Univ. 4:1-81.—Distribution of *Didelphis v. virginiana* includes all of Oklahoma. Systematics, Paleontology, Distribution, Behavior, Genetics, Morphology. Didelphidae.

379 **Gardner, M. C.** 1948. Another Oklahoma armadillo. J. Mammal. 29:76.—*Dasypus novemcinctus* in Wichita Mountains Refuge, Comanche County. Distribution. Dasypodidae.

380 **Garner, G. W.** 1976. Mortality of white-tailed deer fawns in the Wichita Mountains, Comanche County, Oklahoma. Ph.D. Dissertation, Oklahoma State Univ., Stillwater. 113 pp.—*Odocoileus virginianus* fawns in Comanche County had high mortality rate, primarily due to predation by *Canis latrans* and *Felis rufus.* Natural history, Be-

havior, Ecology, Populations, Wildlife management. Canidae, Felidae, Cervidae.

381 **Garner, G. W., and J. A. Morrison.** 1977. Diurnal range and movements of young white-tailed deer fawns in southwestern Oklahoma. Proc. Annu. Conf. Southeast. Assoc. Fish Wildl. Agencies 31:126-133.—*Odocoileus virginianus* fawns in Comanche County have larger diurnal ranges and higher mortality rates than in other areas. Natural history, Behavior, Ecology, Populations. Cervidae.

382 **Garner, G. W., and J. A. Morrison.** 1980. Observations of interspecific behavior between predators and white-tailed deer in southwestern Oklahoma. J. Mammal. 61:126-130.—*Odocoileus virginianus* response to *Canis latrans* and *Felis rufus* dependent on sex and reproductive condition. Behavior. Canidae, Felidae, Cervidae.

383 **Garner, G. W., J. A. Morrison, and J. C. Lewis.** 1978. Mortality of white-tailed deer fawns in the Wichita Mountains, Oklahoma. Proc. Annu. Conf. Southeast. Assoc. Fish Wildl. Agencies 30:493-506.—*Odocoileus virginianus* fawns in Comanche County suffer high mortality, principally from *Canis latrans* and *Felis rufus.* Natural history, Ecology, Populations, Parasites, Wildlife management. Canidae, Felidae, Cervidae.

384 **Garner, G. W., J. Powell, and J. A. Morrison.** 1979. Vegetative composition surrounding daytime bedsites of white-tailed deer fawns in southwestern Oklahoma. Proc. Annu. Conf. Southeast. Assoc. Fish Wildl. Agencies 33:259-266.—Daytime bedsites of *Odocoileus virginianus* fawns in Comanche County primarily in grassy areas. Natural history, Behavior, Ecology. Cervidae.

Garner, G. W., *see* 110.

385 **Garretson, M. S.** 1934. A short history of the American bison. American Bison Society, New York. 66 pp.—Includes brief discussion of herd of *Bison bison* at Wichita Mountains Wildlife Refuge, Comanche County. History, Distribution, Natural history. Bovidae.

386 **Garretson, M. S.** 1938. The American bison. The story of its extermination as a wild species and its restoration under federal protection. New York Zoological Society, New York. 254 pp.—*Bison bison.* Systematics, Paleontology, History, Distribution, Natural history, Behavior, Ecology, Parasites, Morphology, Economics, Wildlife management, Popular. Bovidae.

Geary, D., *see* 1119, 1120.

387 **Geluso, K. N.** 1971. Habitat distribution of *Peromyscus* in the Black Mesa region of Oklahoma. J. Mammal. 52:605-607.—*Peromyscus truei, P. difficilis, P. boylii, P. leucopus* and *P. maniculatus* each found in more than one habitat. Each abundant only in certain areas. Distribution, Ecology. Cricetidae.

388 **Geluso, K. N.** 1972. Western spotted skunk in Oklahoma. Southwest. Nat. 16:457-458.—*Spilogale gracilis* in Cimarron County. Systematics, Distribution, Ecology. Mustelidae.

389 **Geluso, K. N.** 1978. Urine concentrating ability and renal structure of insectivorous bats. J. Mammal. 59:312-323.—*Myotis yumanensis, M. velifer, M. lucifugus, M. leibii, Lasionycteris noctivagans, Pipistrellus hesperus, Eptesicus fuscus, Plecotus townsendii, Tadarida brasiliensis, Antrozous pallidus,* and *Lasiurus cinereus.* Most animals from New Mexico, but *M. velifer* taken in southeastern Oklahoma. Distribution, Ecology, Morphology, Physiology. Vespertilionidae, Molossidae.

Genoways, H. H., *see* 242, 391, 392, 477, 586, 637, 643, 644, 645.

390 **George, S. B.** 1980. Distribution and taxonomic status of *Blarina hylophaga* Elliot. M.S. Thesis, Fort Hays State Univ., Hays, Kansas. —Southwestern (including Oklahoma exclusive of the southeastern corner) populations of *Blarina carolinensis* assigned to name *B. hylophaga.* Systematics, Distribution, Genetics. Soricidae.

391 **George, S. B., J. R. Choate, and H. H. Genoways.** 1981. Distribution and taxonomic status of *Blarina hylophaga* Elliot (Insectivora: Soricidae). Ann. Carnegie Mus. 50:493-513.—All Oklahoma short-tailed shrews referred to *Blarina hylophaga.* Systematics, Distribution. Soricidae.

392 **George, S. B., H. H. Genoways, J. R. Choate, and R. J. Baker.** 1982. Karyotypic relationships within the short-tailed shrews, genus *Blarina.* J. Mammal. 63:639-645.—Includes specimens from Murray and Garvin counties. Blarina from extreme southeastern Oklahoma allocated to *B. carolinensis;* all others within Oklahoma to *B. hylophaga.* Systematics, Distribution, Genetics. Soricidae.

393 **Gidley, J. W.** 1926. Explorations of a Pleistocene spring-deposit in Oklahoma. Smithson. Misc. Collect. 78(1):27-28.—Fossil site in Kiowa County contains horse, mammoth, mastodon, and mylodon bones. Paleontology. Mammalia.

394 **Gier, H. T., S. M. Kruckenberg, and R. J. Marler.** 1978. Parasites and diseases of coyotes. Pages 37-71 *in* Coyotes: Biology, behavior, and management (M. Beckoff, ed.). Academic Press, New York.

384 pp.—Includes information on a number of diseases and parasites that have been found in Oklahoma coyotes (*Canis latrans*). Parasites. Canidae.

Gier, H. T., *see* 810, 811.

395 **Gilbert, C. M.** 1979. Mammoth hunters: The Domebo site. Univ. Okla. Stovall Mus. Sci. Hist. and Okla. Archeol. Surv., Prehistoric People of Okla. 4:1-27.—Sites in Caddo and Kiowa counties yielded remains of mammoths. Paleontology, Popular. Elephantidae.

396 **Gilbert, C. M.** 1980. Oklahoma prehistory. Univ. Okla. Stovall Mus. and Okla. Archeol. Surv., Norman. 80 pp.—Mentions a number of mammals used by prehistoric Oklahomans. Paleontology, Popular. Mammalia.

Gilbert, D. L., *see* 957.

397 **Gillette, D. D., and C. E. Ray.** 1981. Glyptodonts of North America. Smithson. Contrib. Paleobiol. 40:1-255.—Specimens (two broken teeth and carapace) of *Glyptotherium arizonae* cited from Tillman County. Comments on Meade's (1953) mistaken contention that these probably represented a separate genus. Systematics, Paleontology, Distribution. Glyptodontidae.

Gilliam, C. O., *see* 627.

Gilliland, A. R., *see* 1108.

Gilstrap, R. L., *see* 455.

398 **Glass, B. P.** 1947. Geographic variation in *Perognathus hispidus.* J. Mammal. 28:174-179.—*Perognathus* (=*Chaetodipus*) *hispidus spilotus* in eastern and central Oklahoma; *P. h. paradoxus* in western Oklahoma including Panhandle. Systematics, Distribution. Heteromyidae.

399 **Glass, B. P.** 1949. Report on the mammals of the Black Mesa region, Oklahoma. Proc. Okla. Acad. Sci. 30:26-30.—Accounts of 32 mammal species in Cimarron County. Distribution, Ecology. Mammalia.

400 **Glass, B. P.** 1949. Reproduction in the pine vole, *Pitymys nemoralis.* J. Mammal. 30:72-73.—*Pitymys* (=*Microtus*) *nemoralis* (=*pinetorum*) pregnant in January and February. Distribution, Natural history. Cricetidae.

401 **Glass, B. P.** 1951. Ecological factors affecting distribution and speciation of pocket gophers in Oklahoma. Ph.D. Dissertation, Oklahoma State Univ., Stillwater. 78 pp.—*Geomys bursarius* occurs

statewide, while *Cratogeomys castanops* found in Cimarron, Beaver, and Texas counties. Systematics, Paleontology, Distribution, Ecology, Morphology, Bibliographies. Geomyidae.

402 **Glass, B. P.** 1951. A key to the skulls of North American mammals. Burgess Publ. Co., Minneapolis. 52 pp.—Morphology. Mammalia.

403 **Glass, B. P.** 1952. Factors affecting the survival of the plains muskrat, *Ondatra zibethica cinnamomina,* in Oklahoma. J. Wildl. Manage. 16:484-491.—*Ondatra zibethica* (=*zibethicus*). Construction of impoundments has helped muskrats. Suitability for muskrats depends on plant life, water turbidity, and stable water levels. Damage to earthen dams by muskrats can be minimized by using proper construction methods. Distribution, Natural history, Behavior, Ecology, Populations, Economics. Cricetidae.

404 **Glass, B. P.** 1953. A second *Notiosorex* from Oklahoma. J. Mammal. 34:118.—*Notiosorex crawfordi crawfordi.* Skeletal elements recovered from barn owl pellet in Cimarron County. Distribution, Ecology, Morphology. Soricidae.

405 **Glass, B. P.** 1953. Variation in the lower incisors of the Mexican freetail bat, *Tadarida mexicana* (Saussure). Proc. Okla. Acad. Sci. 34:73-74.—Many specimens of *Tadarida mexicana* (=*brasiliensis*) missing one or both lower third incisors, or the tooth is smaller than normal. Morphology. Molossidae.

406 **Glass, B. P.** 1954. Aberrant coloration in *Tadarida mexicana.* Am. Midl. Nat. 52:400-402.—Several instances of partial albinism reported for *Tadarida mexicana* (=*brasiliensis*) from Oklahoma. Natural history. Molossidae.

407 **Glass, B. P.** 1955. New records for bats in Oklahoma. J. Mammal. 36:128.—*Myotis grisescens, M. sodalis, M. velifer, Pipistrellus s. subflavus, Eptesicus f. fuscus, Corynorhinus rafinesquei* (=*Plecotus townsendii*) *pallescens,* and *Tadarida mexicana* (=*brasiliensis*). Distribution, Natural history. Vespertilionidae, Molossidae.

408 **Glass, B. P.** 1956. Effectiveness of Japanese mist nets for securing bats in temperate latitudes. Southwest. Nat. 1:136-138.—*Lasiurus borealis, Pipistrellus subflavus, Eptesicus fuscus, Myotis keenii, M. grisescens, M. velifer, M. subulatus, Antrozous bunkeri* (=*pallidus*), *Tadarida mexicana* (=*brasiliensis*), and *Corynorhinus rafinesquei* (=*Plecotus townsendii*). Payne, Cimarron, and Comanche counties. Distribution, Ecology, Behavior, Techniques. Vespertilionidae, Molossidae.

409 **Glass, B. P.** 1956. Status of the kit fox, *Vulpes velox*, in the High Plains. Proc. Okla. Acad. Sci. 37:162-163.—Study includes specimens of *Vulpes velox* (=*macrotis*) at Oklahoma State University Museum and No-Mans-Land Museum at Goodwell. Texas, Woodward, and Woods counties, Oklahoma, as well as three Texas counties. History, Distribution, Natural history. Canidae.

410 **Glass, B. P.** 1958. Recovery of rabies virus from the Mexican freetail bat in Oklahoma. Proc. Okla. Acad. Sci. 39:83-84.—*Antrozous pallidus, Corynorhinus rafinesquei* (=*Plecotus townsendii*), *Eptesicus fuscus, Lasiurus borealis, L. cinereus, Myotis austroriparius, M. grisescens, M. keeni* (=*keenii*), *M. velifer, M. yumanensis, Nycticeius humeralis, Pipistrellus subflavus, P. hesperus,* and *Tadarida brasiliensis.* Results of rabies tests. Parasites. Vespertilionidae, Molossidae.

411 **Glass, B. P.** 1958. Returns of Mexican freetail bats banded in Oklahoma. J. Mammal. 39:435-436.—*Tadarida brasiliensis mexicana* has southward migration during late fall. Distribution, Natural history, Ecology. Molossidae.

412 **Glass, B. P.** 1958. The Seminole bat in Oklahoma. J. Mammal. 39:587.—*Lasiurus seminolus* in McCurtain County. Distribution. Vespertilionidae.

413 **Glass, B. P.** 1959. Additional returns from free-tailed bats banded in Oklahoma. J. Mammal. 40:542-545.—*Tadarida brasiliensis mexicana.* Distribution, Natural history. Molossidae.

414 **Glass, B. P.** 1960. The taxonomic status of Oklahoma beavers, *Castor canadensis.* Southwest. Nat. 5:21-24.—Study includes *Castor canadensis texensis* and *C. c. missouriensis* from Atoka, Pushmataha, Pawnee, Bryan, Beaver, Caddo, Canadian, Cherokee, Cotton, Ellis, Oklahoma, Pittsburg, and Texas counties, Oklahoma, as well as Kansas and Texas. Systematics, Distribution, Ecology, Morphology. Castoridae.

415 **Glass, B. P.** 1961. Two noteworthy records of bats for Oklahoma. Southwest. Nat. 6:200-201.—*Plecotus townsendii ingens, P. t. pallescens,* and *Lasionycteris noctivagans.* Adair and Payne counties. Distribution, Natural history, Ecology. Vespertilionidae.

416 **Glass, B. P.** 1965. Some notes on reproduction in the red bat, *Lasiurus borealis.* Proc. Okla. Acad. Sci. 46:40-41.—*Lasiurus borealis* in Bryan and LeFlore counties. Natural history, Ecology. Vespertilionidae.

417 **Glass, B. P.** 1971. The type locality of *Dipodomys ordii richardsoni.* Southwest. Nat. 15:497-499.—Type locality of *Dipodops richardsoni*

(=*Dipodomys ordii richardsoni*) in Cimarron County. Also mentions *Bison bison* and *Neotoma micropus canescens.* Systematics, History, Distribution, Ecology. Heteromyidae, Cricetidae, Bovidae.

418 **Glass, B. P.** 1973. A key to the skulls of North American mammals. Second edition. Oklahoma State Univ. Press, Stillwater. 59 pp.—Morphology. Mammalia.

419 **Glass, B. P.** 1982. Seasonal movement of Mexican freetail bats *Tadarida brasiliensis mexicana* banded in the Great Plains. Southwest. Nat. 27:127-133.—Summary of migratory information for *Tadarida brasiliensis* from northwestern Oklahoma caves. Distribution, Natural history, Behavior, Populations. Molossidae.

420 **Glass, B. P., and A. H. [A. F.] Halloran.** 1960. Status and distribution of the red fox, (*Vulpes vulpes*) in Oklahoma. Southwest. Nat. 5:71-74.—Study includes specimens from Cherokee, Bryan, Payne, Adair, Okmulgee, Pushmataha, Marshall, Stephens, Atoka, Canadian, Choctaw, Comanche, Pontotoc, Seminole, Wagoner, Woods, Johnston, Lincoln, McCurtain, Osage, and Pawnee counties. Systematics, Distribution, Morphology, Wildlife management. Canidae.

421 **Glass, B. P., and A. F. Halloran.** 1961. The small mammals of the Wichita Mountains Wildlife Refuge, Oklahoma. J. Mammal. 42:234-239.—Excluding carnivores and ungulates, 34 species recorded for Wichita Mountains Wildlife Refuge, Comanche County. Distribution, Ecology. Mammalia.

422 **Glass, B. P., and W. R. Hanson.** 1952. A feral coatimundi in northwestern Okla. J. Mammal. 33:108-109.—*Nasua narica* (=*nasua*) taken in Woodward County. Natural history, Ecology. Procyonidae.

423 **Glass, B. P., and S. R. Humphrey.** 1971. Distribution of the evening bat *Nycticeius humeralis* in Oklahoma. Southwest. Nat. 15:399-400.—*Nycticeius humeralis* in Tulsa, Payne, Murray, McCurtain, Adair, and LeFlore counties. Distribution, Ecology. Vespertilionidae.

424 **Glass, B. P., and R. C. Morse.** 1959. A new pipistrel from Oklahoma and Texas. J. Mammal. 40:531-534.—*Pipistrellus hesperus oklahomae.* Description of new subspecies. Type specimen from Greer County. Systematics, Morphology. Vespertilionidae.

425 **Glass, B. P., and C. S. Rouk.** 1970. Comparative gastric histology of five North and Central American bats. J. Mammal. 51:455-472.—Specimens from Oklahoma and Mexico. *Tadarida brasiliensis, Myotis velifer,* and *Antrozous pallidus.* Morphology. Vespertilionidae, Molossidae.

426 **Glass, B. P., and C. M. Ward.** 1959. Bats of the genus *Myotis* from Oklahoma. J. Mammal. 40:194-201.—*Myotis lucifugus, M. yumanensis, M. austroriparius, M. grisescens, M. velifer, M. keenii, M. sodalis,* and *M. subulatus.* Distribution, Morphology. Vespertilionidae.

Glass, B. P., *see* 87, 97, 502, 503, 646, 813.

Glass, G. E., *see* 122.

427 **Glenn, B. L., R. E. Rolley, and A. A. Kocan.** 1982. Cytauxzoon-like piroplasms in erythrocytes of wild-trapped bobcats in Oklahoma. J. Am. Vet. Med. Assoc. 181:1251-1253.—*Lynx rufus.* Parasites. Felidae.

Glenn, B. L., *see* 680, 681.

428 **Goertz, J. W.** 1961. A new marginal record for the fulvous harvest mouse in Oklahoma. Proc. Okla. Acad. Sci. 42:124-127.—*Reithrodontomys fulvescens aurantius, R. f. intermedius,* and *R. f. laceyi.* Oklahoma records for *R. fulvescens.* Distribution, Natural history, Ecology, Populations, Morphology. Cricetidae.

429 **Goertz, J. W.** 1962. The influence of habitat quality upon density of cotton rat populations. Ph.D. Dissertation, Oklahoma State Univ., Stillwater. 81 pp.—Payne County study of *Sigmodon hispidus* habitat requirements. Behavior, Ecology, Populations, Techniques. Cricetidae.

430 **Goertz, J. W.** 1962. An opossum-titmouse incident. Wilson Bull. 74:189-190.—Tufted titmouse in Payne County collected hair for nest from two *Didelphis virginiana.* Natural history. Didelphidae.

431 **Goertz, J. W.** 1962. Some biological notes on the plains harvest mouse. Proc. Okla. Acad. Sci. 43:123-125.—*Reithrodontomys montanus* in Payne County. Natural history, Ecology, Populations, Morphology. Cricetidae.

432 **Goertz, J. W.** 1962. A visceral abnormality in the cotton rat. J. Mammal. 43:269.—*Sigmodon hispidus* from Payne County had splenic, mesenteric, and caecal deformations apparently due to old injury. Natural history, Morphology. Cricetidae.

433 **Goertz, J. W.** 1964. The influence of habitat quality upon density of cotton rat populations. Ecol. Monogr. 34:359-381.—*Sigmodon hispidus texianus* in Payne County. Live-trapping and snap-trapping. Habitat preferences, home range, distance traveled, disappearance rates, and population fluctuations. Distribution, Natural history, Behavior, Ecology, Populations. Cricetidae.

434 **Goertz, J. W.** 1965. Adrenal and spleen weights in cotton rats in relation to age, sex, and season. J. Mammal. 46:699-700.—*Sigmodon hispidus.* Adrenal glands of females heavier than those of males on average. Spleen weight is highly variable. Morphology, Physiology. Cricetidae.

435 **Goertz, J. W.** 1965. Reproductive variation in cotton rats. Am. Midl. Nat. 74:329-340.— *Sigmodon hispidus* populations from Payne County potentially irruptive throughout year. Natural history, Ecology, Populations, Morphology. Cricetidae.

436 **Goertz, J. W.** 1965. Sex, age, and weight variation in cotton rats. J. Mammal. 46:471-477.—*Sigmodon hispidus* populations studied for two years. Sex ratio, weight and age distribution variation attributable to various factors. Populations. Cricetidae.

437 **Goertz, J. W.** 1966. Incidence of warbles in some Oklahoma rodents. Am. Midl. Nat. 75:242-245.—Botfly larvae found in *Perognathus, Sigmodon, Neotoma,* and *Peromyscus,* but absent in *Rattus, Mus, Reithrodontomys,* and *Microtus* in Payne County. Natural history, Parasites. Heteromyidae, Cricetidae, Muridae.

438 **Goertz, J. W.** 1970. An ecological study of *Neotoma floridana* in Oklahoma. J. Mammal. 51:94-104.—Habitat, sex ratios, reproductive conditions, and movements of *Neotoma floridana* studied. Distribution, Natural history, Ecology, Populations. Cricetidae.

439 **Goertz, J. W.** 1971. An ecological study of *Microtus pinetorum* in Oklahoma. Am. Midl. Nat. 86:1-12.—*Microtus pinetorum, Sigmodon hispidus, Neotoma floridana,* and *Peromyscus* species in Payne County. Natural history, Ecology, Populations, Morphology. Cricetidae.

440 **Goldman, E. A.** 1910. Revision of the wood rats of the genus *Neotoma.* N. Am. Fauna 31:1-124.—Distributions of *Neotoma floridana baileyi, N. f. attwateri, N. m. micropus,* and *N. m. canescens* include portions of Oklahoma. Systematics, Distribution, Natural history, Behavior, Ecology, Morphology, Economics. Cricetidae.

441 **Goldman, E. A.** 1914. The rice rats of North America (genus *Oryzomys*). N. Am. Fauna 43:1-100.—*Oryzomys palustris texensis* occurs in Oklahoma. Systematics, History, Distribution, Natural history, Morphology, Economics. Cricetidae.

442 **Goldman, E. A.** 1937. The wolves of North America. J. Mammal. 18: 37-45.—Historical distributions of *Canis lupus* and *C. rufus* included Oklahoma. Systematics, Distribution. Canidae.

443 **Goldman, E. A.** 1941. A new western subspecies of golden mouse. Proc. Biol. Soc. Wash. 54:189-191.—Range of *Peromyscus nuttalli* (=*Ochrotomys nuttalli) flammeus* includes Oklahoma, with specimens from Redland, Sequoyah County. Systematics, Distribution, Morphology. Cricetidae.

444 **Goldman, E. A.** 1950. Raccoons of North and Middle America. N. Am. Fauna 60:1-153.—*Procyon lotor hirtus* and *P. l. fuscipes* listed for Oklahoma. Systematics, History, Distribution, Natural history, Behavior, Ecology, Morphology, Economics. Procyonidae.

Goldman, E. A., *see* 831, 1199, 1200.

Good, N. E., *see* 888.

Goodrum, P., *see* 361.

445 **Gould, C. N.** 1927. My friend *Glyptodon* and his pals. My Okla. 1(8): 32, 33, 36, 52.—Discussion of fossils including a glyptodon, elephant, camel, horse, and ground sloth found at Tillman County site. Paleontology, Popular. Glyptodontidae, Mammalia.

446 **Gould, C. N.** 1928. The fossil *Glyptodon* in the Frederick gravel beds. Proc. Okla. Acad. Sci. 8:148-150.—Fossils of elephant, camel, glyptodon, ground sloth, horse, tapir and pecarry. Paleontology. Glyptodontidae, Mammalia.

447 **Gould, C. N.** 1929. Fossil bones and artifacts at Frederick. Proc. Okla. Acad. Sci. 9:90-92.—Fossils of horses, *Megalonyx, Mylodon, Glyptodon,* tapir, pecarry, llama, camel, mastodons, and elephants found in gravel beds in Tillman County. Paleontology. Mammalia.

448 **Graham, G.** 1938. Animal outlaws. Second edition. Gideon Graham, Collinsville, Oklahoma. 256 pp.—Wildlife anecodotes. Distribution, Natural history, Popular. Mammalia.

449 **Grant, W. E., and E. C. Birney.** 1979. Small mammal community structure in North American grasslands. J. Mammal. 60:23-36.—Osage County site supports high small mammal biomass. Ecology, Populations. Soricidae, Heteromyidae, Cricetidae, Muridae.

450 **Grant, W. E., E. C. Birney, N. R. French, and V. M. Swift.** 1982. Structure and productivity of grassland small mammal communities relating to grazing-induced changes in vegetative cover. J. Mammal. 63:248-260.—Study of rodent population densities includes tallgrass site in Osage County. Ecology, Populations. Soricidae, Sciuridae.

Grant, W. E., *see* 144, 373.

451 **Grau, H. J.** 1981. Kin recognition in *Peromyscus leucopus.* M.S. Thesis, Univ. Oklahoma, Norman. 41 pp.—Study of *Peromyscus leucopus* indicates ability to recognize close kin without previous encounter. Behavior, Genetics. Cricetidae.

452 **Grau, H. J.** 1982. Kin recognition in white-footed deermice (*Peromyscus leucopus*). Anim. Behav. 30:497-505.—*Peromyscus leucopus* show behavioral discrimination between unfamiliar kin and non-kin. Study of laboratory stocks originated from animals captured near Norman. Behavior, Genetics. Cricetidae.

453 **Gray, G. G., and C. D. Simpson.** 1980. *Ammotragus lervia.* Mamm. Species 144:1-7.—The Barbary sheep was introduced in the southwestern United States. Free-ranging distribution included southwestern corner and western Panhandle of Oklahoma. Summary of information in literature. Systematics, Paleontology, Distribution, Natural history, Behavior, Ecology, Populations, Parasites, Genetics, Morphology, Physiology. Bovidae.

454 **Green, P. M.** 1964. Density of population as a regulating factor in the reproductive potential of *Sigmodon hispidus.* Ph.D. Dissertation, Oklahoma State Univ., Stillwater. 101 pp.—Payne County *Sigmodon hispidus* cease breeding earlier in fall under high population densities. Natural history, Populations, Morphology, Physiology. Cricetidae.

Green, P. M., *see* 351.

455 **Greer, J. K., and R. L. Gilstrap.** 1970. Vertebrate remains in barn owl pellets. Bull. Okla. Ornithol. Soc. 3:23-29.—*Tyto alba* in Canadian County preys on at least 11 species of mammals. Natural history, Ecology. Soricidae, Talpidae, Leporidae, Geomyidae, Heteromyidae, Cricetidae.

Greer, J. K., *see* 1085, 1086.

Griffith, M. E., *see* 1164.

456 **Grigsby, E. M.** 1980. The gray bat, *Myotis grisescens,* in the southwest portion of the Ozark plateau: Movement patterns, maternity colonies, hibernacula and philopatry. Ph.D. Dissertation, Univ. Oklahoma, Norman. 107 pp.—Study of *Myotis grisescens* includes localities in Cherokee, Delaware, Adair, and Ottawa counties. Distribution, Natural history, Behavior, Ecology, Populations. Vespertilionidae.

457 **Grinnell, G. B.** 1929. Pronghorn antelope. J. Mammal. 10:135-141.—*Antilocapra americana* stocked on Wichita Mountains Refuge,

Comanche County, but none survived. Wildlife management. Antilocapridae.

Grossenheider, R. P., *see* 192, 193.

458 **Grubitz, G.** 1963. The social behavior of the thirteen-lined ground squirrel (*Citellus tridecemlineatus*). M.S. Thesis, Univ. Oklahoma, Norman. 57 pp.—*Citellus* (=*Spermophilus*) *tridecemlineatus.* Cleveland County. Some information on similar studies on other members of the family. Partial summary of physiological studies of *S. tridecemlineatus.* Behavior, Physiology. Sciuridae.

Grzybowski, J. A., *see* 960.

459 **Guberlet, J. E.** 1923. Parasites of dogs and cats of Oklahoma. Proc. Okla. Acad. Sci. 3:71-78.—External parasites, protozoa, trematodes, tapeworms, and roundworms of feral *Canis familiaris* and *Felis catus.* Parasites. Canidae, Felidae.

Guilday, J. E., *see* 576.

Gunier, W. J., *see* 324, 857.

460 **Haines, H., G. Chichester, and H. F. Landreth, Jr.** 1977. Blood respiratory properties of *Bison bison.* Resp. Physiol. 30:305-310.—Evaluated blood taken from bison of the Wichita Wildlife Refuge, Cache, Oklahoma. Physiology. Bovidae.

461 **Haines, H., C. Ciskowski, and V. Harms.** 1973. Acclimation to chronic water restriction in the wild house mouse *Mus musculus.* Physiol. Zool. 46:110-125.—Specimens from colony at University of Oklahoma. Colony stocked from specimens collected near Norman and Clinton. Once established, acclimated state was maintained; food consumption not different in acclimated mice. Information on water income, urinary water loss, urine concentration and insensible evaporation. Physiology. Muridae.

462 **Haines, H., T. M. McKenna, and J. E. Melton.** 1978. Body fluid distribution in wild *Mus musculus* acclimated to water restriction. Am. J. Physiol. 235:R237-R242.—House mice used were from laboratory bred mice that originated from animals captured in Norman, Oklahoma. Physiology. Muridae.

463 **Haines, H., and K. Schmidt-Nielsen.** 1967. Water deprivation in wild house mice. Physiol. Zool. 40:424-431.—Includes *Mus musculus* from colony at University of Oklahoma stocked with individuals collected in Cleveland County. No differences found in urine-concentrating ability that were related to geographical origin of colonies. Average

maximum osmolality of urine was 5.0 osmol. *Mus musculus* can tolerate water deprivation better than nondesert rodents. Ecology, Physiology. Muridae.

Haines, H., *see* 208, 322, 775, 806.

464 **Hair, J. A.** 1968. Ticks can kill! Outdoor Okla. 24(11):2-7.—Ticks responsible for death of many *Odocoileus virginianus* fawns. Natural history, Parasites, Popular. Cervidae.

465 **Hair, J. A., D. E. Howell, C. E. Rogers, and J. Fletcher.** 1969. Occurrence of the pharyngeal bot *Cephenemyia jellisoni* in Oklahoma white-tailed deer, *Odocoileus virginianus.* Ann. Entomol. Soc. Am. 62:1208-1210.—Pharyngeal bot fly larvae found in majority of *Odocoileus virginianus* examined in Adair, Cherokee, and Pittsburg counties. Parasites. Cervidae.

Hair, J. A., *see* 170, 258, 865, 866, 977.

Haley, J. D., *see* 504.

466 **Hall, E. R.** 1936. Mustelid mammals from the Pleistocene of North America with systematic notes on some Recent members of the genus *Mustela, Taxidea,* and *Mephitis.* Carnegie Inst. Wash. Publ. 473:41-119.—Distributions of various Pleistocene and Recent mustelids include portions of Oklahoma. Systematics, Paleontology, Distribution. Mustelidae.

467 **Hall, E. R.** 1946. A new genus of American Pliocene badger, with remarks on the relationship of badgers of the Northern Hemisphere. Carnegie Inst. Wash. Publ. 551:9-23.—Includes discussion of *Pliotaxidea nevadensis* from Texas County. Systematics, Paleontology. Mustelidae.

468 **Hall, E. R.** 1951. American weasels. Univ. Kans. Publ. Mus. Nat. Hist. 4:1-466.—*Mustela frenata primulina, M. f. neomexicana,* and *M. f. texensis* recorded in Oklahoma. Systematics, Paleontology, History, Distribution, Natural history, Behavior, Ecology, Populations, Morphology. Mustelidae.

469 **Hall, E. R.** 1951. A synopsis of the North American Lagomorpha. Univ. Kans. Publ. Mus. Nat. Hist. 5:119-202.—*Sylvilagus floridanus, S. audubonii, S. aquaticus,* and *Lepus californicus* in Oklahoma. Keys to families and genera of Lagomorpha. Systematics, Distribution, Natural history, Behavior, Ecology, Parasites, Morphology. Leporidae.

470 **Hall, E. R.** 1954. A new subspecies of pocket mouse from Kansas. Univ. Kans. Publ. Mus. Nat. Hist. 7:587-590.—Distribution of *Perognathus flavescens cockrumi* extends at least into Dewey County, Oklahoma. Systematics, Distribution, Morphology. Heteromyidae.

471 **Hall, E. R.** 1955. Handbook of mammals of Kansas. Univ. Kans. Mus. Nat. Hist. Misc. Publ. 7:1-303.—Records of mammals along northern border of Oklahoma. Distribution, Natural history. Mammalia.

472 **Hall, E. R.** 1965. Names of species of North American mammals north of Mexico. Univ. Kans. Mus. Nat. Hist. Misc. Publ. 43:1-16.—Checklist of scientific and common names of all mammals north of Mexico. Systematics. Mammalia.

473 **Hall, E. R.** 1981. The mammals of North America. Second edition. John Wiley and Sons, New York. 2 vols. 1,175 + 10 + 90 pp.—Systematics, Paleontology, History, Distribution, Morphology, Techniques. Mammalia.

474 **Hall, E. R., S. Anderson, J. K. Jones, Jr., and R. L. Packard.** 1957. Vernacular names for North American mammals north of Mexico. Univ. Kans. Mus. Nat. Hist. Misc. Publ. 14:1-16.—Checklist of scientific and common names of all mammals north of Mexico. Systematics. Mammalia.

475 **Hall, E. R., and E. L. Cockrum.** 1953. A synopsis of the North American microtine rodents. Univ. Kans. Publ. Mus. Nat. Hist. 5:373-498.—*Microtus ochrogaster, M. pinetorum, Ondatra zibethicus,* and *Synaptomys cooperi* occur in Oklahoma. Systematics, Distribution, Populations, Parasites, Morphology, Economics. Cricetidae.

476 **Hall, E. R., and W. W. Dalquest.** 1950. A synopsis of the American bats of the genus *Pipistrellus.* Univ. Kans. Publ. Mus. Nat. Hist. 1: 591-602.—Includes Oklahoma specimens. Systematics, Distribution, Morphology. Vespertilionidae.

477 **Hall, E. R., and H. H. Genoways.** 1970. Taxonomy of the *Neotoma albigula*-group of woodrats in central Mexico. J. Mammal. 51:504-516. —*Neotoma albigula warreni* occurs in Cimarron County. Systematics, Distribution, Morphology. Cricetidae.

478 **Hall, E. R., and K. R. Kelson.** 1952. Comments on the taxonomy and geographic distribution of some North American marsupials, insectivores, and carnivores. Univ. Kans. Publ. Mus. Nat. Hist. 5:319-341.—Specimens of *Blarina brevicauda* from Tulsa County assigned to

B. b. carolinensis (=*B. hylophaga*). Systematics, Distribution. Soricidae.

479 **Hall, E. R., and K. R. Kelson.** 1959. The mammals of North America. Ronald Press Co., New York. 2 vol. 1,083 pp.— Systematics, Paleontology, History, Distribution, Morphology, Techniques. Mammalia.

480 **Hall, S.** 1952. Fossil tusk found near Rosedale, Oklahoma. Okla. Anthropol. Soc. Newsl. 1(6):3.—Tusk of mammoth or mastodon. Paleontology. Mastodontidae, Elephantidae.

Hall, S., *see* 723, 560.

481 **Hall, S. A.** 1982. Late Holocene paleoecology of the Southern Plains. Quat. Res. (N.Y.) 17:391-407.—Analysis of pollen and land snails from rockshelter sites in northeastern Oklahoma indicate that 2000-1000 yr B.P. was moister than today. Comments on mammal fauna. Paleontology, Distribution, Ecology. Cricetidae, Bovidae, Mammalia.

482 **Halloran, A. F.** 1956. Last bears of the Wichitas. Okla. Game Fish News 12(7):3,11.—*Ursus americanus* in Comanche County. Account for the layman. Past records of black bears. History, Distribution, Popular. Ursidae.

483 **Halloran, A. F.** 1956. Wichita antelope on comeback trail. Okla. Game Fish News 12(12):6-7.—History of *Antilocapra americana* introductions and management at Wichita Mountains Wildlife Refuge, Comanche County. History, Natural history, Ecology, Wildlife management, Popular. Antilocapridae.

484 **Halloran, A. F.** 1957. The bison returned to the plains. Anim. Kingdom 60:130-135.—History of *Bison bison* herd at Wichita Mountains Wildlife Refuge, Comanche County. History, Ecology, Populations. Bovidae.

485 **Halloran, A. F.** 1957. Keeping wildlife refuge grass and buffalo-longhorn herds in balance. Cattleman 44:86-87.—Wichita Mountains Wildlife Refuge, in Comanche County, maintains range by disposing of surplus *Bison bison* and *Bos taurus.* History, Ecology, Wildlife management. Bovidae.

486 **Halloran, A. F.** 1957. Lion in Oklahoma—Past and present. Okla. Game Fish News 13(2):6-8.—*Felis concolor.* For the layman. Past accounts in Oklahoma. History, Distribution, Popular. Felidae.

487 **Halloran, A. F.** 1957. Live and dressed weights of American bison. J. Mammal. 38:139.—*Bison bison* from Comanche County. Natural history, Populations, Morphology. Bovidae.

488 **Halloran, A. F.** 1957. Longhorn cattle weights and horn measurements. Proc. Okla. Acad. Sci. 38:135-136.—*Bos taurus* in Comanche County. Morphology, Wildlife management. Bovidae.

489 **Halloran, A. F.** 1958. Black red wolves. Okla. Wildl. 14(4):6-8.—Historical distribution of *Canis niger* (=*rufus*) in southeastern Oklahoma, while that of *Canis latrans* in western Oklahoma. History, Distribution, Ecology, Popular. Canidae.

490 **Halloran, A. F.** 1960. American bison weights and measurements from the Wichita Mountains Wildlife Refuge. Proc. Okla. Acad. Sci. 41:212-218.—*Bison bison* in Comanche County. History, Natural history, Ecology, Populations, Morphology, Wildlife management. Bovidae.

491 **Halloran, A. F.** 1961. Additional longhorn cattle management records from Wichita Mountains Wildlife Refuge. Proc. Okla. Acad. Sci. 42:268-271.—*Bos taurus* in Comanche County. History, Natural history, Populations, Morphology, Wildlife management. Bovidae.

492 **Halloran, A. F.** 1962. History of the Wichita Mountains Wildlife Refuge elk herd. Proc. Okla. Acad. Sci. 43:229-232.—*Cervus canadensis* (=*elaphus*) *nelsoni, C. c.* (=*elaphus*) *merriami* in Comanche County. History, Distribution, Populations, Morphology. Cervidae.

493 **Halloran, A. F.** 1963. A melanistic coyote from Oklahoma. Southwest. Nat. 8:48-49.—*Canis latrans* in Comanche County. Specimen deposited in University of Oklahoma Stovall Museum (now Okla. Mus. Nat. Hist.). Morphology. Canidae.

494 **Halloran, A. F.** 1963. Specimens of the black bear (*Euarctos americanus*) from the Wichita Mountains of Oklahoma. Southwest. Nat. 8:174.—*Euarctos* (=*Ursus*) *americanus* in Comanche County. Distribution. Ursidae.

495 **Halloran, A. F.** 1965. Texas longhorn cattle weights and measurements from Wichita Mountains Wildlife Refuge, Oklahoma. Proc. Okla. Acad. Sci. 46:228-232.—*Bos taurus* in Comanche County. Natural history, Morphology. Bovidae.

496 **Halloran, A. F.** 1966. The size and weight of elk from the Wichita Wildlife Refuge, Oklahoma. Proc. Okla. Acad. Sci. 47:406-413.—

Cervus canadensis (=*elaphus*) *nelsoni* from Comanche County. History, Natural history, Ecology, Populations. Cervidae.

497 **Halloran, A. F.** 1967. Deer live-trapping history on the Wichita Mountains Wildlife Refuge, Oklahoma. Proc. Okla. Acad. Sci. 48:205-207.—*Odocoileus virginianus.* Delaware and Comanche counties. Map of release records. Distribution, Economics, Wildlife management. Cervidae.

498 **Halloran, A. F.** 1968. Bison (Bovidae) productivity on the Wichita Mountains Wildlife Refuge, Oklahoma. Southwest. Nat. 13:23-26.—*Bison bison* in Comanche County. Natural history, Ecology, Populations. Bovidae.

499 **Halloran, A. F.** 1970. An experimental introduction of bighorn sheep into the Wichita Mountains of Oklahoma. Great Plains J. 9:85-90.—History of introduction and failure of *Ovis canadensis* in Wichita Mountains Wildlife Refuge, Comanche County. History, Parasites. Bovidae.

500 **Halloran, A. F.** 1972. The black-tailed prairie dog: Yesterday and today. Great Plains J. 11:138-144.—Includes brief discussion of *Cynomys ludovicianus* populations in Oklahoma. History, Natural history, Populations. Sciuridae.

501 **Halloran, A. F.** 1975. Wildlife sanctuary of the forests and plains mammals of the Wichitas. Great Plains J. 14:174-209.—Annotated list of mammals known to occur in Wichita Mountains Wildlife Refuge, Comanche County. Distribution, Natural history. Mammalia.

502 **Halloran, A. F., and B. P. Glass.** 1959. The carnivores and ungulates of the Wichita Mountains Wildlife Refuge, Oklahoma. J. Mammal. 40:360-370.—History, Distribution, Natural history, Ecology, Populations. Mammalia.

503 **Halloran, A. F., and B. P. Glass.** 1963. Additional mammal notes from the Wichita Mountains region of Oklahoma. Proc. Okla. Acad. Sci. 44:56-58.—Accounts of nine species from Comanche County. Distribution, Ecology. Mammalia.

504 **Halloran, A. F., and J. D. Haley.** 1962. A bibliography of the Wichita Mountains of Oklahoma. Great Plains J. 1(2):1-26.—Includes lists of popular, general, historical and scientific references to Wichita Mountains Wildlife Refuge, Comanche County. Bibliographies. Mammalia.

505 **Halloran, A. F., and C. A. Shrader.** 1960. Longhorn cattle management on Wichita Mountains Wildlife Refuge. J. Wildl. Manage. 24: 191-196.—Description of morphologic and behavioral characteristics of longhorn cattle, *Bos taurus,* as well as management techniques on Wichita Mountains Refuge. Systematics, History, Behavior, Genetics, Morphology, Wildlife management. Bovidae.

Halloran, A. F., *see* 420, 421.

506 **Handley, C. O., Jr.** 1956. A third specimen of the shrew *Notiosorex* from Oklahoma. J. Mammal. 37:278.—*Notiosorex crawfordi* in Comanche County. Distribution, Morphology. Soricidae.

507 **Handley, C. O., Jr.** 1959. A revision of American bats of the genera *Euderma* and *Plecotus.* Proc. U.S. Natl. Mus. 110:95-246.—Distributions of *Plecotus townsendii pallescens, P. t. ingens,* and *P. rafinesquii* include portions of Oklahoma. Systematics, Paleontology, Distribution, Ecology, Morphology. Vespertilionidae.

Hansen, M. F., *see* 835.

Hanson, W. R., *see* 422.

508 **Harden, W. D.** 1972. Predation by hawks on bats at Vickery bat cave. Bull. Okla. Ornithol. Soc. 5:4-5.—*Tadarida brasiliensis* preyed on by *Buteo swainsoni* and *Falco sparverius. Buteo jamaicensis* made unsuccessful attempt. Natural history, Ecology. Molossidae.

Hardisty, R. M., *see* 209.

Harlow, R. F., *see* 354.

Harms, V., *see* 461.

Haroldson, K. J., *see* 374.

509 **Harriman, A. E.** 1973. Self-selection of diet in northern grasshopper mice (*Onychomys leucogaster*). Am. Midl. Nat. 90:97-106.—*Onychomys leucogaster breviauritus* from Harper County. Appetites exhibited during dietary self-selection reflect physiologic needs. Distribution, Natural history, Ecology, Behavior. Cricetidae.

510 **Harris, A. H.** 1974. *Myotis yumanensis* in interior southwestern North America, with comments on *Myotis lucifugus.* J. Mammal. 55:589-607. —Description of *Myotis lucifugus relictus,* a new subspecies. Systematics, History, Distribution, Populations, Morphology. Vespertilionidae.

Harris, A. H., *see* 356.

511 **Harrison, J. A.** 1981. A review of the extinct wolverine, *Plesiogulo* (Carnivora: Mustelidae), from North America. Smithson. Contrib. Paleobiol. 46:1-27.—Specimen referenced from Texas County. Systematics, Paleontology, Distribution. Mustelidae.

Hart, B. J., *see* 1204, 1205.

512 **Hart, E. B.** 1971. Karyology and evolution of the plains pocket gopher, *Geomys bursarius.* Ph.D. Dissertation, Univ. Oklahoma, Norman. 111 pp.—*Geomys bursarius major, G. b. breviceps,* and *G. b. bursarius* occur in Oklahoma. Systematics, Paleontology, Distribution, Genetics, Techniques, Bibliographies. Geomyidae.

513 **Hart, E. B.** 1978. Karyology and evolution of the plains pocket gopher, *Geomys bursarius.* Occas. Pap. Mus. Nat. Hist. Univ. Kans. 71:1-20.—Includes Oklahoma specimens. Systematics, History, Distribution, Ecology, Genetics. Geomyidae.

514 **Hartley, J. D., and A. F. Miller.** 1977. Archeological investigations at the Bryson-Paddock site, an early contact period site on the Southern Plains. Okla. River Basin Surv. Archaeol. Site Rep. 32:1-264.—Investigators identified 3,207 of 34,626 bone fragments recovered. Bison (*Bison bison*) and white-tailed deer (*Odocoileus virginianus*) were species most frequently encountered. Paleontology. Cervidae, Bovidae, Mammalia.

515 **Hartman, S. E.** 1979. Geographic variation analysis of 11 populations of *Dipodomys ordii* (Heteromyidae) using nonmetric cranial traits. M.S. Thesis, Univ. Oklahoma, Norman. 35 pp.—Includes eight populations of *Dipodomys ordii richardsoni* and *D. o. oklahomae* from Oklahoma. Systematics, Distribution, Morphology. Heteromyidae.

516 **Hartman, S. E.** 1980. Geographic variation analysis of *Dipodomys ordii* using nonmetric cranial traits. J. Mammal. 61:436-448.—Eleven *Dipodomys ordii* populations (eight from Oklahoma) sampled for nonmetric cranial traits. Few intertrait correlations found. Distribution, Genetics, Morphology. Heteromyidae.

517 **Harvey, M. J.** 1976. Endangered Chiroptera of the southeastern United States. Proc. Annu. Conf. Southeast. Assoc. Game Fish Comm. 29:429-433.—Discussion of status of *Myotis sodalis, M. grisescens,* and *Plecotus townsendii ingens.* Distributions include Oklahoma. Distribution, Natural history, Ecology, Populations, Wildlife management. Vespertilionidae.

518 **Harvey, M. J.** 1980. Status of the endangered bats *Myotis sodalis, M. grisescens,* and *Plecotus townsendii ingens* in southern Ozarks. Proc.

Int. Bat Res. Conf. 5:221-223.—Discussion of status of populations of *Myotis sodalis, M. grisescens,* and *Plecotus townsendii ingens* in Arkansas and Oklahoma. Distribution, Populations, Wildlife management. Vespertilionidae.

519 **Hassien, F. D.** 1976. A search for black-footed ferrets in the Oklahoma Panhandle and adjacent area and an ecological study of black-tailed prairie dogs in Texas County, Oklahoma. M.S. Thesis, Oklahoma State Univ., Stillwater. 111 pp.—Study in Texas County showed that *Cynomys ludovicianus* abundant, but *Mustela nigripes* rare or extirpated. History, Distribution, Ecology, Populations, Wildlife management. Sciuridae, Mustelidae.

Hassien, F. D., *see* 716, 717.

520 **Hatcher, R.** 1981. Distribution and population trends. Okla. Dep. Wildl. Cons. Job 02, Study 05, W-82-R-20. 5 pp.—Data on distributions and population trends of Oklahoma furbearers. Distribution, Populations. Didelphidae, Canidae, Procyonidae, Mustelidae, Felidae.

521 **Hatcher, R.** 1981. Furbearer harvest. Okla. Dep. Wildl. Cons. Job 07, Study 05, W-82-R-20. 5 pp.—Summary of Oklahoma furbearers taken in 1980-81. Populations, Economics, Wildlife management. Mammalia.

522 **Hatcher, R.** 1981. Furbearer productivity. Okla. Dep. Wildl. Cons. Job 06, Study 05, W-82-R-20. 7 pp.—Skulls and jaws of harvested Oklahoma furbearers collected in addition to portions of carcasses of coyotes, bobcats, and raccoons. Populations, Economics, Wildlife management. Canidae, Procyonidae, Mustelidae, Felidae.

523 **Hatcher, R.** 1981. Hunter-trapper survey. Okla. Dep. Wildl. Cons. Job 04, Study 05, W-82-R-20. 9 pp.—Questionaires (about 2,000) to furbearer hunters and trappers concerning species pursued and taken, as well as methods used. Distribution, Wildlife management. Didelphidae, Castoridae, Cricetidae, Canidae, Procyonidae, Mustelidae, Felidae.

524 **Hatcher, R.** 1981. Reference data and literature compilation. Okla. Dep. Wildl. Cons. Job 01, Study 05, W-82-R-20. 2 pp.—Data and literature on Oklahoma fur harvest. History, Distribution, Wildlife management, Bibliographies. Didelphidae, Canidae, Procyonidae, Mustelidae, Felidae.

525 **Hatcher, R.** 1981. Stock raccoons? Outdoor Okla. 37:27-29.—Discussed past attempts to stocking raccoons (*Procyon lotor*) and con-

cluded that such efforts typically unsuccessful. History, Populations, Economics, Wildlife management, Popular. Procyonidae.

526 **Hatcher, R.** 1982. Distribution and population trends. Okla. Dep. Wildl. Cons. Job 02, Study 05, W-82-R-21. 2 pp.—Furbearer population indices based on scent-post surveys. Data for 1977-81. Populations, Wildlife management. Mammalia.

527 **Hatcher, R.** 1982. Furbearer harvest. Okla. Dep. Wildl. Cons. Job 07, Study 05, W-82-R-21. 5 pp.—Summary of furbearer harvest data for 1981-82. Populations, Economics, Wildlife management. Mammalia.

528 **Hatcher, R.** 1983. Distribution and population trends. Okla. Dep. Wildl. Cons. Job 02, Study 05, W-82-R-22. 25 pp.—Scent-station data from 70 counties and 9 wildlife management areas used to assess populations of furbearers. Populations, Wildlife management. Didelphidae, Canidae, Procyonidae, Mustelidae, Felidae.

529 **Hatcher, R.** 1983. Furbearer harvest. Okla. Dep. Wildl. Cons. Job 07, Study 05, W-82-R-22. 5 pp.—Summary of fur-dealer reports for 1982-83, with comparison to 1981-82. Average pelt prices included. Populations, Economics, Wildlife management. Mammalia.

530 **Hatcher, R.** 1983. Furbearer productivity. Okla. Dep. Wildl. Cons. Job 06, Study 05, W-82-R-22. 3 pp.—Determined age, sex, and average litter size of Oklahoma badgers, bobcats, coyotes, gray foxes, minks and raccoons. Natural history, Ecology. Canidae, Procyonidae, Mustelidae, Felidae.

531 **Hatcher, R.** 1983. Furbearer status report. Okla. Dep. Wildl. Cons. Job 09, Study 05, W-82-R-22. 49 pp.—Summarizes available data concerning fur harvest, pelt prices, and number of trappers. Includes badger, beaver, bobcat, coyote, fox, mink, muskrat, nutria, opossum, raccoon, ringtails, skunks, weasel, and wolf. Populations, Economics, Wildlife management. Mammalia.

532 **Hatcher, R.** 1984. Distribution and population trends. Okla. Dep. Wildl. Cons. Job 02, Study 05, W-82-R-23. 2 pp.—Furbearer population indices based on scent-post surveys. Populations, Wildlife management. Mammalia.

533 **Hatcher, R.** 1984. Furbearer harvest. Okla. Dep. Wildl. Cons. Job 07, Study 05, W-82-R-23. 5 pp.—Summary of fur-dealer reports. Populations, Economics, Wildlife management. Mammalia.

534 **Hatcher, R.** 1984. Furbearer productivity. Okla. Dep. Wildl. Cons. Job 06, Study 05, W-82-R-23. 3 pp.—Determined age, sex, and average litter size of Oklahoma furbearers. Natural history, Ecology. Mammalia.

Hatcher, R., *see* 287.

535 **Hatcher, R. T.** 1979. Status of the red fox in Oklahoma and comparison of 3 furbearer survey techniques. M.S. Thesis, Oklahoma State Univ., Stillwater. 72 pp.—*Vulpes vulpes* distributed statewide except Panhandle. Distributions and abundances also assessed for *Canis latrans, Urocyon cinereoargenteus,* and *Felis rufus.* Distribution, Populations, Techniques, Wildlife management. Canidae, Felidae.

536 **Hatcher, R. T.** 1982. Distribution and status of red foxes (Canidae) in Oklahoma. Southwest. Nat. 27:183-186.—*Vulpes vulpes* occurs statewide, with greatest numbers in oak-hickory forest of northeastern Oklahoma. Distribution, Populations, Techniques, Wildlife management. Canidae.

537 **Hatcher, R. T.** 1984. River otters in Oklahoma. Proc. Okla. Acad. Sci. 64:17-19.—Summary of occurrences of *Lutra canadensis* in Oklahoma. Evaluates reported sightings and other records. History, Distribution, Wildlife management. Mustelidae.

538 **Hatcher, R. T.** 1985. Distribution and population trends. Okla. Dep. Wildl. Cons. Job 01, Study 03, W-82-R-24. 9 pp.—Furbearer population indices based on scent-post surveys. Populations, Wildlife management. Mammalia.

539 **Hatcher, R. T.** 1985. Furbearer productivity. Okla. Dep. Wildl. Cons. Job 04, Study 03, W-82-R-24. 23 pp.—Determined age, sex, and average litter size of Oklahoma furbearers. Natural history, Ecology. Mammalia.

540 **Hatcher, R. T.** 1986. Distribution and population trends. Okla. Dep. Wildl. Cons. Job 01, Study 03, W-82-R-25. 9 pp.—Furbearer population indices based on scent-post surveys. Populations, Wildlife management. Mammalia.

541 **Hatcher, R. T.** 1986. Furbearer harvest. Okla. Dep. Wildl. Cons. Job 05, Study 03, W-82-R-25. 7 pp.—Summary of fur-dealer reports. Populations, Economics, Wildlife management. Mammalia.

542 **Hatcher, R. T.** 1986. Furbearer productivity. Okla. Dep. Wildl. Cons. Job 04, Study 03, W-82-R-25. 12 pp.—Determined age, sex, and

average litter size of Oklahoma furbearers. Natural history, Ecology. Mammalia.

543 **Hatcher, R. T.** 1986. Oklahoma's fur harvest: A fifty year review. Okla. Dep. Wildl. Cons. Fin. W-087-R-25. 58 pp.—First compilation of data on Oklahoma's fur harvest since 1944 report. Based on biennial and annual reports produced by Department of Wildlife Conservation. Data from fur dealers. Includes short history of furbearer statutes. History, Populations, Economics, Wildlife management. Didelphidae, Cricetidae, Myocastoridae, Canidae, Ursidae, Procyonidae, Mustelidae, Felidae.

544 **Hatcher, R. T., and J. H. Shaw.** 1981. A comparison of three indices to furbearer populations. Wildl. Soc. Bull. 9:153-156.—Study in Kiowa, Canadian, Atoka, Delaware, and Latimer counties compares effectiveness of three censusing methods for *Canis latrans, Lynx* (=*Felis*) *rufus, Urocyon cinereoargenteus,* and *Vulpes vulpes.* Distribution, Populations, Techniques, Wildlife management. Canidae, Felidae.

545 **Hatcher, R. T., and G. W. Wigtail.** 1985. Fates of red foxes released into southeastern Oklahoma. Proc. Annu. Conf. Southeast. Assoc. Fish Wildl. Agencies 39:321-325.—In 1981, eight red foxes (*Vulpes vulpes*) released in September in LeFlore County. Most died within two months. Radio transmitters used. Distribution, Ecology, Wildlife management. Canidae.

546 **Hay, O. P.** 1920. Descriptions of some Pleistocene vertebrates found in the United States. Bovidae. Proc. U.S. Natl. Mus. 58:83-146.—Fossil deposits in spring near Afton, Oklahoma. Type locality of *Equus holmesi, Camelops mitidus,* and *Symbos promptus.* Paleontology. Equidae, Camelidae, Bovidae, Mammalia.

547 **Hay, O. P.** 1924. The Pleistocene of the middle region of North America and its vertebrated animals. Carnegie Inst. Wash. Publ. 322A:1-385.—Complete listing of Pleistocene mammals, with localities, known to have been found in Oklahoma. Paleontology. Mammalia.

548 **Hay, O. P.** 1930. Second bibliography and catalogue of the fossil Vertebrata of North America. Carnegie Inst. Wash. Publ. 390(1):1-916, (2):1-1074.—Extensive bibliography arranged and indexed taxonomically, not geographically. Paleontology, Bibliographies. Mammalia.

549 **Hay, O. P., and H. J. Cook.** 1928. Preliminary descriptions of fossil mammals recently discovered in Oklahoma, Texas and New Mexico.

Proc. Colo. Mus. Nat. Hist. 8:33.—*Elephas haroldcooki* found at Tillman County site. Paleontology. Elephantidae.

550 **Hay, O. P., and H. J. Cook.** 1930. Fossil vertebrates collected near, or in association with, human artifacts at localities near Colorado, Texas; Frederick, Oklahoma; and Folsom, New Mexico. Proc. Colo. Mus. Nat. Hist. 9:4-40.—Fossil site in Tillman County contains specimens of mammals. Type locality for *Equus achates, Lama hollomani,* and *Gomphotherium priestleyi.* Paleontology. Gomphotheriidae, Equidae, Camelidae, Mammalia.

551 **Hays, H. A.** 1954. Mammals of grassland and wooded ravine in central Oklahoma, their ecology and faunal relationships. Ph.D. Dissertation, Univ. Oklahoma, Norman. 102 pp.—McClain County study of habits and habitats of small- and medium-sized mammals. Distribution, Natural history, Behavior, Ecology, Populations, Techniques. Mammalia.

552 **Hays, H. A.** 1958. The distribution and movement of small mammals in central Oklahoma. J. Mammal. 39:235-244.—*Peromyscus maniculatus, Perognathus* (=*Chaetodipus*) *hispidus* and *Sigmodon hispidus* most common animals on grassland. *Peromyscus leucopus* and *Neotoma floridana* most common in wooded area. Most mammals remained in one vegetative type. Distribution, Natural history, Ecology, Populations. Heteromyidae, Cricetidae.

553 **Hays, H. A.** 1958. The effect of microclimate on the distribution of small mammals in a tall-grass prairie plot. Trans. Kans. Acad. Sci. 61: 40-63.—Extensive discussion of status and habitat requirements of 25 species of small mammals in tallgrass area in McClain County. Distribution, Ecology. Mammalia.

554 **Hayward, B. J.** 1970. The natural history of the cave bat *Myotis velifer.* Western New Mexico Univ. Res. Sci. 1:1-74.—Distribution, Natural history. Vespertilionidae.

555 **Heaney, L. R., and R. M. Timm.** 1983. Relationships of pocket gophers of the genus *Geomys* from the central and northern Great Plains. Univ. Kans. Mus. Nat. Hist. Misc. Publ. 74:1-59.—Morphometric analyses employing skin and skull measures to assess geographic variation and interspecific associations. Includes detailed assessment of *Geomys breviceps* and *G. lutescens* from Oklahoma. Bacula, karyotypes, and parasites also analyzed. Systematics, Distribution, Natural history, Parasites, Genetics, Morphology. Geomyidae.

556 **Heard, W. R.** 1958. Cotton rats, *Sigmodon hispidus,* as food of channel catfish, *Ictalurus punctatus.* Proc. Okla. Acad. Sci. 39:200-201.—Several gill-netted channel catfish in Payne County had eaten *Sigmodon hispidus.* Natural history, Ecology. Cricetidae.

557 **Hedrick, R. D.** 1978. Impact of channelization and associated land use changes on wildlife habitat. M.S. Thesis, Oklahoma State Univ., Stillwater. 46 pp.—Study assessed impact of stream channelization on wildlife in Carter, Grady, Garvin, Murray, and Stephens counties. Ecology, Wildlife management. Mammalia.

558 **Heist, T. A.** 1970. Biosystematics of *Neotoma floridana* and *Neotoma micropus* in Oklahoma. M.S. Thesis, Oklahoma State Univ., Stillwater. 44 pp.—Study verifies *Neotoma micropus* and *N. floridana* as valid, distinguishable species. Systematics, Distribution, Genetics, Morphology. Cricetidae.

559 **Hellack, J. J.** 1974. Mammals of Oklahoma. Pages 140-146 *in* Field guide to Oklahoma (P. G. Risser, ed.). Oklahoma Biological Survey, Norman, Oklahoma. 160 pp.—Checklist of Oklahoma mammals and their distributions within state. Distribution, Natural history. Mammalia.

Hellack, J. J., *see* 1132.

Henderson, F. R., *see* 241.

560 **Henry, D. O., B. H. Butler, and S. A. Hall.** 1979. The Late Prehistoric human ecology of Birch Creek Valley, northeastern Oklahoma. Plains Anthropol. 24:207-238.—Sixteen mammal species recovered, with white-tailed deer (*Odocoileus virginianus*) being most abundant. Paleontology. Cervidae, Mammalia.

561 **Hepworth, J. L.** 1966. Hematology of *Sigmodon hispidus*: Average parameters compared with those under induced stresses. Ph.D. Dissertation, Oklahoma State Univ., Stillwater. 71 pp.—Payne County study of *Sigmodon hispidus* shows different blood characteristic values for stressed and non-stressed individuals. Behavior, Physiology, Techniques. Cricetidae.

Hepworth, J. L., *see* 138.

562 **Herd, J., and D. Curry.** 1977. Upland game investigations. Okla. Dep. Wildl. Cons. Job 01, Study 01, W-82-R-16. 13 pp.—Includes census information of squirrel population trends 1972-77. Populations, Wildlife management. Sciuridae.

563 **Herd, J., and D. Curry.** 1978. Upland game investigations. Okla. Dep. Wildl. Cons. Job 01, Study 01, W-82-R-17. 10 pp.—Update of census information, including that for squirrel surveys. Populations, Wildlife management. Sciuridae.

564 **Hermanson, J. W., and T. J. O'Shea.** 1983. *Antrozous pallidus.* Mamm. Species 213:1-8.—Summary of information in literature. *Antrozous pallidus* occurs in western one-third of Oklahoma. Systematics, Paleontology, History, Distribution, Natural history, Behavior, Ecology, Populations, Parasites, Genetics, Morphology, Economics, Bibliographies. Vespertilionidae.

565 **Herreid, C. F., II.** 1963. Temperature regulation of Mexican free-tailed bats in cave habitats. J. Mammal. 44:560-573.—*Tadarida brasiliensis* cluster temperatures from Reed Cave, Harmon County. Physiology. Molossidae.

566 **Herrmann, J. A.** 1952. Morphological variation in the Texas races of the wood-mouse, *Peromyscus leucopus.* M. A. Thesis, Univ. Texas, Austin. 36 pp.—Geographic variation study of *Peromyscus leucopus* includes specimens from Bryan, Pushmataha, Harmon, and Jackson counties. Systematics, Distribution, Morphology. Cricetidae.

567 **Hesse, C. J.** 1936. A Pliocene vertebrate fauna from Optima, Oklahoma. Univ. Calif. Publ. Geol. Sci. 24:57-69.—Texas County sites yielded 13 genera of mammals. Paleontology. Mylagaulidae, Canidae, Mustelidae, Camelidae, Equidae, Rhinocerotidae.

568 **Hesse, C. J.** 1938. Lower Pliocene vertebrate fossils from the Ogallala formation (Laverne zone) of Beaver County, Oklahoma. Carnegie Inst. Wash. Publ. 476:47-72.—Sites in Beaver County yielded remains of rodents, carnivores, horses, deer, and camels. Paleontology. Castoridae, Procyonidae, Equidae, Cervidae, Camelidae, Mammalia.

569 **Hibbard, C. W.** 1933. A revised check list of Kansas mammals. Trans. Kans. Acad. Sci. 36:230-249.—Discussion of taxonomy and status of 109 species and subspecies of Kansas mammals. Systematics, Distribution. Mammalia.

570 **Hibbard, C. W.** 1934. The occurrence of *Erethizon epixanthum bruneri* and *Mustela nigripes* in Kansas. J. Mammal. 15:70-71.—*Mustela nigripes* in Texas County. Distribution. Mustelidae.

571 **Hibbard, C. W.** 1937. *Peromyscus leucopus aridulus* in Kansas. J. Mammal. 18:101-102.—Includes speculation on occurrence of *Peromyscus leucopus aridulus* in Oklahoma. Distribution. Cricetidae.

572 **Hibbard, C. W.** 1944. A check-list of Kansas mammals, 1943. Trans. Kans. Acad. Sci. 47:61-88.—Taxonomy and status of Kansas mammals. Systematics, Distribution. Mammalia.

573 **Hibbard, C. W.** 1954. A new Pliocene vertebrate fauna from Oklahoma. Pap. Mich. Acad. Sci. Arts Lett. 39:339-359.—Type locality in Beaver County for *Buisnictis schoffi, Citellus dotti,* and *Pliogeomys buisi.* Paleontology. Geomyidae, Sciuridae, Mustelidae, Mammalia.

574 **Hibbard, C. W.** 1958. Summary of North American Pleistocene mammalian faunas. Pap. Mich. Acad. Sci. Arts Lett. 43:1-32.—Lists 29 species of mammals known from Illinoian and Kansan faunas in Oklahoma. Paleontology, Bibliographies. Mammalia.

575 **Hibbard, C. W.** 1970. Pleistocene mammalian local faunas from the Great Plains and Central Lowland provinces of the United States. Pages 395-433 *in* Pleistocene and Recent environments of the central Great Plains (W. Dort, Jr., and J. K. Jones, Jr., eds.). Univ. Press Kansas, Lawrence. 433 pp.—Includes some Oklahoma paleontological localities. Paleontology. Mammalia.

576 **Hibbard, C. W., D. E. Ray, D. E. Savage, D. W. Taylor, and J. E. Guilday.** 1965. Quaternary mammals of North America. Pages 509-525 *in* The Quaternary of the United States (H. E. Wright, Jr., and D. G. Frey, eds.). Princeton Univ. Press, Princeton, New Jersey. 922 pp.—Discussion refers to four Oklahoma Pleistocene localities. Paleontology, Bibliographies. Mammalia.

577 **Hibbard, C. W., and D. W. Taylor.** 1960. Two Late Pleistocene faunas from southwestern Kansas. Contr. Mus. Paleontol. Univ. Mich. 16:1-223.—Includes brief discussion and mammal checklists for Berends local fauna, Beaver County, and Bar M local fauna, Harper County. Paleontology. Mammalia.

Hibbard, C. W., *see* 912, 1050, 1072.

578 **Hicks, L.** 1974. Wildlife portrait: The prairie dog. Outdoor Okla. 30(5):20.—*Cynomys ludovicianus.* Account for the layman. Distribution, Natural history, Behavior, Ecology, Morphology, Popular. Sciuridae.

579 **Hicks, L.** 1975. Wildlife portrait: The beaver. Outdoor Okla. 31(10):20.—*Castor canadensis.* Account for the layman. Distribution, Natural history, Ecology, Morphology, Popular. Castoridae.

580 **Hicks, L.** 1975. Wildlife portrait: The white foot. Outdoor Okla. 31(8):20.—*Peromyscus leucopus.* Account for the layman. Distribu-

tion, Natural history, Behavior, Ecology, Populations, Morphology, Popular. Cricetidae.

581 **Hicks, L.** 1975. Wildlife portrait: The woodchuck. Outdoor Okla. 31(7):20.—*Marmota monax.* Account for the layman. Distribution, Natural history, Behavior, Ecology, Morphology, Economics, Popular. Sciuridae.

582 **Hillman, C. N., and T. W. Clark.** 1980. *Mustela nigripes.* Mamm. Species 126:1-3.—Distribution of black-footed ferret, *Mustela nigripes,* includes western Oklahoma. Summary of information in literature. Systematics, Paleontology, History, Distribution, Natural history, Behavior, Ecology, Populations, Parasites, Morphology, Bibliographies. Mustelidae.

Hoar, D., *see* 991.

583 **Hoar, R.** 1981. Squirrel population trends. Okla. Dep. Wildl. Cons. Job 04, Study 01, W-82-R-20. 2 pp.—Monitoring fox squirrel (*Sciurus niger*) and gray squirrel (*S. carolinensis*) populations. Distribution, Populations. Sciuridae.

Hoar, R., *see* 993, 1118.

Hockman, J. G., *see* 230.

584 **Hoditschek, B., and T. L. Best.** 1983. Reproductive biology of Ord's kangaroo rat (*Dipodomys ordii*) in Oklahoma. J. Mammal. 64:121-127.—Assessment of age, environmental, and seasonal parameters in *Dipodomys ordii* in central Oklahoma. Natural history, Morphology. Heteromyidae.

Hoditschek, B., *see* 129, 130.

585 **Hoffmann, R. S., and E. C. Birney.** 1972. Ecological studies of small mammal populations at the Cottonwood and Osage sites, 1971. U.S. IBP (Int. Biol. Program) Grassland Biome Tech. Rep. 187:1-45. Colorado State Univ., Fort Collins.—Small mammal populations assessed on Osage County site, with *Microtus ochrogaster* and *Sigmodon hispidus* being dominant species. Ecology, Populations. Soricidae, Leporidae, Heteromyidae, Cricetidae, Muridae.

586 **Hoffmann, R. S., J. K. Jones, Jr., and H. H. Genoways.** 1971. Small mammal survey on the Bison, Bridger, Cottonwood, Dickinson, and Osage sites. U.S. IBP (Int. Biol. Program) Grassland Biome Tech. Rep. 109:1-69. Colorado State Univ., Fort Collins.—Small mammal populations assessed on Osage County site, with *Microtus ochrogaster*

being dominant species. Ecology, Populations. Soricidae, Sciuridae, Cricetidae, Mammalia.

Hoffmann, R. S., *see* 122, 645.

587 **Hoffmeister, D. F.** 1951. A taxonomic and evolutionary study of the piñon mouse, *Peromyscus truei.* Ill. Biol. Monogr. 21(4):1-104.—*Peromyscus t. truei* occurs in Cimarron County. Systematics, Distribution, Morphology. Cricetidae.

588 **Hoffmeister, D. F.** 1981. *Peromyscus truei.* Mamm. Species 161:1-5.—Summary of information in literature. Range of piñon mouse, *Peromyscus truei truei,* extends into Cimarron County. Systematics, Paleontology, History, Distribution, Natural history, Behavior, Ecology, Genetics, Morphology, Physiology, Bibliographies. Cricetidae.

589 **Hoffmeister, D. F., and L. de la Torre.** 1961. Geographic variation in the mouse *Peromyscus difficilis.* J. Mammal. 42:1-13.—*Peromyscus nasutus* considered to be subspecies of *P. difficilis.* Cimarron County. Systematics, Distribution, Morphology. Cricetidae.

590 **Hofman, J.** 1971. Bison bone flesher from Caddo County. Okla. Anthropol. Soc. Newsl. 21(2):5-7.—Notched bison-bone flesher found. Paleontology. Bovidae.

591 **Hofman, J. L.** 1974. Bone fleshers in Oklahoma. Bull. Okla. Anthropol. Soc. 23:169-175.—Thirteen bone fleshers known for Oklahoma. Most made from bison (*Bison bison*) bones. Paleontology. Bovidae.

592 **Holle, D. G.** 1977. Diet and general availability of prey of the coyote (*Canis latrans*) at Wichita Mountains National Wildlife Refuge, Oklahoma. M.S. Thesis, Oklahoma State Univ., Stillwater. 59 pp.—Fecal analysis of *Canis latrans* in Comanche County showed that primary food items were rodents and rabbits. Small mammal abundances estimated by trapping. Natural history, Behavior, Ecology, Populations, Wildlife management. Leporidae, Cricetidae, Canidae, Mammalia.

593 **Holle, D. G.** 1978. Food habits of coyotes in an area of high fawn mortality. Proc. Okla. Acad. Sci. 58:11-15.—Coyotes (*Canis latrans*) in Comanche County eat primarily rodents, in addition to some *Odocoileus virginianus* fawns. Natural history, Ecology. Canidae, Cervidae.

594 **Hollister, N.** 1911. A systematic synopsis of the muskrats. N. Am. Fauna 32:1-147.—*Fiber* (=*Ondatra*) *zibethicus cinnamominus* occurs

in northern Oklahoma. Systematics, Paleontology, History, Distribution, Natural history, Behavior, Ecology, Morphology, Economics. Cricetidae.

595 **Hollister, N.** 1913. Three new subspecies of grasshopper mice. Proc. Biol. Soc. Wash. 26:215-216.—Type locality of *Onychomys leucogaster breviauritus* is Fort Reno, Canadian County. Systematics, Distribution, Morphology. Cricetidae.

596 **Hollister, N.** 1914. A systematic account of the grasshopper mice. Proc. U.S. Natl. Mus. 47:427-490.—Distributions of *Onychomys leucogaster articeps* and *O. l. longipes* include portions of Oklahoma. Systematics, Distribution, Natural history, Morphology. Cricetidae.

597 **Hollister, N.** 1916. A systematic account of the prairie dogs. N. Am. Fauna 40:1-37.—*Cynomys l. ludovicianus* occurs in Oklahoma. Systematics, Distribution, Natural history, Morphology, Economics. Sciuridae.

598 **Holmes, W. H.** 1902. Flint implements and fossil remains from a sulphur spring at Afton, Indian Territory. Am. Anthropol. 4:108-129.—Excavation of spring revealed fossil teeth of mastodons, mammoths, bison, and horses. Paleontology. Bovidae, Equidae, Elephantidae, Mastodontidae.

599 **Holmes, W. H.** 1903. Flint implements and fossil remains from a sulphur spring at Afton, Indian Territory. Rep. U.S. Natl. Mus. 1901:237-252.—Excavation of spring produced bones of mammoths, mastodons, bison, horses, deer, elk, and wolves. Paleontology. Mammalia.

Homan, J. A., *see* 647.

600 **Honacki, J. H., K. E. Kinman, and J. W. Koeppl.** 1982. Mammal species of the world. Allen Press and Association of Systematics Collections, Lawrence, Kansas. 694 pp.—Checklist of all mammal species of world. Systematics, Distribution, Bibliographies. Mammalia.

601 **Honeycutt, R. L.** 1978. Chromosomal and morphological variation in *Geomys bursarius* (Shaw) from Texas and adjacent states with comments on factors influencing distribution. M.S. Thesis, Texas A and M Univ., College Station. 106 pp.—*Geomys bursarius sagittalis* occurs in eastern Oklahoma, while *G. b. major* found in western part of state. Systematics, Distribution, Natural history, Ecology, Genetics, Morphology, Bibliographies. Geomyidae.

602 **Honeycutt, R. L., and D. J. Schmidly.** 1979. Chromosomal and morphological variation in the plains pocket gopher, *Geomys bursarius*, in Texas and adjacent states. Occas. Pap. Mus. Tex. Tech Univ. 58:1-54. —*Geomys bursarius dutcheri* is a synonym of *G. b. sagittalis*. Systematics, Distribution. Geomyidae.

603 **Hood, C. S., L. W. Robbins, R. J. Baker, and H. S. Shellhammer.** 1984. Chromosomal studies and evolutionary relationships of an endangered species, *Reithrodontomys raviventris*. J. Mammal. 65:655-667.— Study includes *Reithrodontomys fulvescens* specimens from McIntosh County. Systematics, Distribution. Cricetidae.

604 **Hooper, E. T.** 1952. Notes on mice of the species *Peromyscus boylii* and *P. pectoralis*. J. Mammal. 33:371-378.—Description of new subspecies, *Peromyscus pectoralis collinus*. Systematics, Distribution, Ecology, Morphology. Cricetidae.

605 **Hooper, E. T.** 1952. A systematic review of the harvest mice (genus *Reithrodontomys*) of Latin America. Misc. Publ. Mus. Zool. Univ. Mich. 77:1-255.—Distributions of *Reithrodontomys fulvescens aurantius* and of *R. montanus* include portions of Oklahoma. Systematics, Distribution, Natural history, Morphology. Cricetidae.

606 **Hoover, J. P.** 1984. Surgical implantation of radiotelemetry devices in American river otters. J. Am. Vet. Med. Assoc. 185:1317-1320.— Devices implanted in river otters (*Lutra canadensis*) to be released in southeastern Oklahoma. Techniques, Wildlife management. Mustelidae.

607 **Hoover, J. P., R. J. Bahr, M. A. Nieves, R. T. Doyle, M. A. Zimmer, and S. E. Lauzon.** 1985. Clinical evaluation and prerelease management of American river otters in the second year of reintroduction study. J. Am. Vet. Med. Assoc. 187:1154-1161.—Information on river otters (*Lutra canadensis*) before reintroduction. Parasites, Physiology, Wildlife management. Mustelidae.

608 **Hoover, J. P., A. E. Castro, and M. A. Nieves.** 1985. Serologic evaluation of vaccinated American river otters. J. Am. Vet. Med. Assoc. 187:1162-1165.—Data on river otters (*Lutra canadensis*) prior to reintroduction. Parasites, Physiology, Wildlife management. Mustelidae.

609 **Hoover, J. P., C. R. Root, and M. A. Zimmer.** 1984. Clinical evaluation of American river otters in a reintroduction study. J. Am. Vet. Med. Assoc. 185:1321-1326.—Information on river otters (*Lutra*

canadensis) before introduction. Parasites, Physiology, Wildlife management. Mustelidae.

Hoppe, K. L., *see* 635.

Horner, N. V., *see* 293.

Hornuff, L., *see* 210, 211.

Hough, D. J., *see* 960.

Howell, A. B., *see* 311.

610 **Howell, A. H.** 1901. Revision of the skunks of the genus *Chinca.* N. Am. Fauna 20:1-47.—Includes specimens of *Chinca* (=*Mephitis*) *mesomelas* (=*mephitis*) from Oklahoma. Systematics, Distribution, Morphology. Mustelidae.

611 **Howell, A. H.** 1906. Revision of the skunks of the genus *Spilogale.* N. Am. Fauna 26:1-55.—*Spilogale interrupta* (=*putorius*) occurs throughout most of Oklahoma. Systematics, History, Distribution, Natural history, Behavior, Ecology, Morphology. Mustelidae.

612 **Howell, A. H.** 1914. Revision of the American harvest mice. N. Am. Fauna 36:1-97.—*Reithrodontomys albescens* (=*montanus*) *griseus, R. fulvescens aurantius,* and *R. f. intermedius* found in Oklahoma. Systematics, History, Distribution, Natural history, Ecology, Morphology, Economics. Cricetidae.

613 **Howell, A. H.** 1915. Revision of the American marmots. N. Am. Fauna 37:1-80.—*Marmota monax monax* occurs in eastern Oklahoma. Systematics, History, Distribution, Natural history, Ecology, Morphology, Economics. Sciuridae.

614 **Howell, A. H.** 1918. Revision of the American flying squirrels. N. Am. Fauna 44:1-64.—Distributions of *Glaucomys v. volans* and *G. v. saturatus* include portions of Oklahoma. Systematics, Distribution, Natural history, Behavior, Ecology, Morphology, Economics. Sciuridae.

615 **Howell, A. H.** 1929. Revision of the American chipmunks. N. Am. Fauna 52:1-157.—*Tamias striatus venustus* and possibly *Eutamias* (=*Tamias*) *q. quadrivittatus* occur in Oklahoma. Systematics, History, Distribution, Natural history, Behavior, Ecology, Morphology, Economics. Sciuridae.

616 **Howell, A. H.** 1938. Revision of the North American ground squirrels with a classification of the North American Sciuridae. N. Am. Fauna 56:1-256.—*Citellus* (=*Spermophilus*) *tridecemlineatus texensis, C.* (=*S.*)

t. arenicola, and *C.* (=*S.*) *spilosoma major* occur in Oklahoma. Systematics, History, Distribution, Natural history, Behavior, Ecology, Morphology, Economics. Sciuridae.

Howell, D. E., *see* 258, 465.

617 **Howell, P.** 1977. The skin game. Outdoor Okla. 33(10):7-11.—Statistics on changing market value of furbearers. Economics, Wildlife management, Popular. Mammalia.

Hubbell, T. H., *see* 165.

Huckaby, D. G., *see* 695.

Hughes, R. C., *see* 1161.

618 **Hughes, W.** 1950. Trapping bobcats in Kay County. Okla. Game Fish News 6(12):12-13.—*Felis rufus* in Kay County. Distribution, Natural history, Popular. Felidae.

Huheey, C. C., *see* 704.

619 **Humphrey, S. R.** 1969. Disturbances and bats. Okla. Underground 2:42-44.—Discusses effects of both summer and winter disturbances on *Plecotus townsendii, Myotis velifer,* and *Tadarida brasiliensis.* Natural history, Populations, Wildlife management. Vespertilionidae, Molossidae.

620 **Humphrey, S. R.** 1974. Zoogeography of the nine-banded armadillo (*Dasypus novemcinctus*) in the United States. BioScience 24:457-462. —Discussion of factors encouraging and restricting range expansion of *Dasypus novemcinctus.* Distribution, Natural history, Ecology. Dasypodidae.

621 **Humphrey, S. R.** 1975. Extension of known range of the desert shrew *Notiosorex crawfordi* in Oklahoma. Am. Midl. Nat. 94:499-500.—*Notiosorex crawfordi* in Woodward County. Examined pellets from barn owl (*Tyto alba*). Records from Oklahoma. Also mentioned *Neotoma micropus.* Distribution, Ecology. Soricidae, Cricetidae.

622 **Humphrey, S. R.** 1975. Nursery roosts and community diversity of Nearctic bats. J. Mammal. 56:321-346.—Compares faunal size with physical structure of environment. Considers distributions in United States. Distribution, Behavior, Ecology, Populations, Physiology. Vespertilionidae, Molossidae.

623 **Humphrey, S. R., and T. H. Kunz.** 1976. Ecology of a Pleistocene relict, the western big-eared bat (*Plecotus townsendii*) in the southern Great Plains. J. Mammal. 57:470-494.—*Plecotus townsendii* in Major,

Woodward and Woods counties, Oklahoma, as well as Barber and Comanche counties, Kansas. History, Distribution, Natural History, Behavior, Ecology, Populations, Physiology. Vespertilionidae.

Humphrey, S. R., *see* 171, 423.

Hunkapillar, T., *see* 151.

Hunsicker, R., *see* 151.

Hunt, O. E., *see* 1057.

Hutchison, V. H., *see* 343.

Inglis, J. M., *see* 771.

624 Jackson, H. H. T. 1915. A review of the American moles. N. Am. Fauna 38:1-100.—*Scalopus aquaticus pulcher, S. a. intermedius,* and *S. a. aereus* occur in Oklahoma. Systematics, History, Distribution, Natural history, Behavior, Ecology, Morphology, Economics. Talpidae.

625 Jackson, H. H. T., and H. E. Warfel. 1933. Notes on the occurrence of mammals in the regions adjacent to the Salt Plains of northwestern Oklahoma. Publ. Okla. Biol. Surv. 5:65-72.—Alfalfa, Major, and Woods counties. Distribution, Ecology. Mammalia.

Jackson, H. H. T., *see* 1201.

626 Jacobs, K. F. 1956. Southwestern Oklahoma game restoration and investigations: Trapping and releasing prairie chicken. Okla. Dep. Wildl. Cons. Job 07, W-065-R-01. 12 pp.—Study mentions several rodent species. Natural history, Ecology, Economics, Wildlife management. Mammalia.

627 Jacobs, K. F., and C. O. Gilliam. 1962. Upland game management and operations: Upland game surveys and post-release observations—exotic species. Okla. Dep. Wildl. Cons. Job 08, W-083-D-01. 22 pp.—Study reports predation by *Canis latrans* and *Lynx* (=*Felis*) *rufus* on exotic game birds. Natural history, Ecology, Economics. Canidae, Felidae.

628 Jacobs, K. F., F. A. James, and G. B. Wint. 1958. Southwestern Oklahoma game restoration and investigations: Evaluation of stubble quail (*Coturnix c. japonica*) releases. Okla. Dep. Wildl. Cons. Job 15, W-065-R-03. 33 pp.—Study includes predation on quail by *Canis familiaris* and *Felis catus. Sigmodon hispidus* also mentioned. Natural history, Ecology, Economics, Wildlife management. Cricetidae, Canidae, Felidae.

629 **Jacobs, K. F., F. A. James, and G. B. Wint.** 1959. Southwestern Oklahoma game restoration and investigations: Evaluation of winter hatched bobwhite quail released in early spring. Okla. Dep. Wildl. Cons. Job 20, W-065-R-04. 13 pp.—Study includes predation on quail by *Canis familiaris, Mephitis mephitis,* and *Felis catus.* Natural history, Ecology, Economics, Wildlife management. Canidae, Mustelidae, Felidae.

630 **Jahoda, J. C.** 1969. Circadian rhythm and lunar periodism in *Onychomys leucogaster breviauritus* (Rodentia; Cricetidae). Ph.D. Dissertation, Oklahoma State Univ., Stillwater. 119 pp.—*Onychomys leucogaster* from Harper County showed increased activity levels under lower light conditions, resulting in daily and monthly activity patterns. Natural history, Behavior, Ecology, Physiology, Techniques. Cricetidae.

631 **Jahoda, J. C.** 1973. The effect of the lunar cycle on the activity pattern of *Onychomys leucogaster breviauritus.* J. Mammal. 54:544-549.—*Onychomys leucogaster* in Harper County show daily and lunar periodicity. Behavior, Ecology. Cricetidae.

James, F. A., *see* 628, 629.

632 **Jameson, E. W., Jr.** 1947. Natural history of the prairie vole (mammalian genus *Microtus*). Univ. Kans. Publ. Mus. Nat. Hist. 1:125-151.—Includes map with Oklahoma distribution. Distribution, Natural history, Behavior, Ecology, Populations, Parasites, Morphology. Cricetidae.

Jamieson, R., *see* 124.

633 **Jenkins, S. H., and P. E. Busher.** 1979. *Castor canadensis.* Mamm. Species 120:1-8.—Distribution of *Castor canadensis* includes all of Oklahoma. Summary of information in literature. Systematics, Paleontology, Distribution, Natural history, Behavior, Ecology, Populations, Parasites, Genetics, Morphology, Physiology, Techniques, Bibliographies. Castoridae.

Jensen, J. F., *see* 1109.

634 **Jobling, B.** 1938. A revision of the species of the genus *Trichobius* (Diptera Acalypterae, Streblidae). Parasitology 30:358-387.—*Trichobius corynorhini* taken from unknown host in Woodward County. Parasites. Vespertilionidae, Molossidae.

635 **John, D. T., and K. L. Hoppe.** 1986. *Trypanosoma cruzi* from wild raccoons in Oklahoma. Am. J. Vet. Res. 47:1056-1059.—*Procyon lotor.* Parasites. Procyonidae.

Johnston, C. S., *see* 1051, 1052, 1053.

636 **Jones, C.** 1977. *Plecotus rafinesquii.* Mamm. Species 69:1-4.—*Plecotus rafinesquii* distribution includes extreme eastern Oklahoma. Summary of information in literature. Systematics, Paleontology, Distribution, Natural history, Behavior, Ecology, Populations, Genetics, Morphology, Physiology, Bibliographies. Vespertilionidae.

Jones, C., *see* 226, 356, 357, 645.

637 **Jones, C. A., J. R. Choate, and H. H. Genoways.** 1984. Phylogeny and paleobiogeography of short-tailed shrews (genus *Blarina*). Pages 56-148 *in* Contributions in Quaternary vertebrate paleontology: A volume in memorial to John E. Guilday (H. H. Genoways and M. R. Dawson, eds.). Carnegie Mus. Nat. Hist. Spec. Publ. No. 8. 538 pp.—Includes Oklahoma specimens of Elliot's short-tailed shrew (*Blarina hylophaga*) as well as fossil shrews from two Oklahoma localities (Doby Springs, Berends). Systematics, Paleontology, Distribution, Morphology. Soricidae.

638 **Jones, G. E.** 1946. The raccoon in Oklahoma. Okla. Game Fish News 2(12):4-5.—*Procyon lotor.* Account for the layman. Distribution, Techniques, Economics, Wildlife management, Popular. Procyonidae.

639 **Jones, G. E.** 1953. Survey of existing beaver range in Oklahoma. Okla. Game Fish Dep. Fed. Aid. Div. Proj. WS1-R2, Completion Rep. 21 pp. Mimeo.—*Castor canadensis.* Distribution, Populations. Castoridae.

640 **Jones, G. E., and J. Mahaffey.** 1952. Beaver survey. Okla. Game Fish News 8(11):8-10.—Assessment indicated small populations of *Castor canadensis* occurring statewide. Distribution, Populations, Wildlife management. Castoridae.

Jones, G. S., *see* 864.

641 **Jones, J. K., Jr., and S. Anderson.** 1959. The eastern harvest mouse, *Reithrodontomys humulis* in Oklahoma. Southwest. Nat. 4:153-154.—*Reithrodontomys humulis, R. fulvescens aurantius, Peromyscus leucopus leucopus, Sigmodon hispidus texianus, Neotoma floridana osagensis,* and *Microtus pinetorum nemoralis.* Latimer County. New

record of *R. humulis* for Oklahoma. Distribution, Ecology, Morphology. Cricetidae.

642 **Jones, J. K., Jr., D. M. Armstrong, and J. R. Choate.** 1986. Guide to mammals of the plains states. Univ. Nebraska Press, Lincoln. 371 pp. —Covers 138 native and 8 introduced species found in North Dakota, South Dakota, Nebraska, Kansas, and Oklahoma. Includes general information as well as keys and distribution maps. Systematics, Distribution, Natural history, Ecology, Morphology. Mammalia.

643 **Jones, J. K., Jr., D. C. Carter, and H. H. Genoways.** 1973. Checklist of North American mammals north of Mexico. Occas. Pap. Mus. Tex. Tech Univ. 12:1-14.—Checklist of scientific and common names of all mammals north of Mexico. Systematics, Bibliographies. Mammalia.

644 **Jones, J. K., Jr., D. C. Carter, and H. H. Genoways.** 1979. Revised checklist of North American mammals north of Mexico, 1979. Occas. Pap. Mus. Tex. Tech Univ. 62:1-15.—Checklist of scientific and common names of all mammals north of Mexico. Systematics, Bibliographies. Mammalia.

645 **Jones, J. K., Jr., D. C. Carter, H. H. Genoways, R. S. Hoffmann, D. W. Rice, and C. Jones.** 1986. Revised checklist of North American mammals north of Mexico, 1986. Occas. Pap. Mus. Tex. Tech Univ. 107:1-22.—Checklist of scientific and common names of all mammals north of Mexico. Systematics, Bibliographies. Mammalia.

646 **Jones, J. K., Jr., and B. P. Glass.** 1960. The short-tailed shrew, *Blarina brevicauda,* in Oklahoma. Southwest. Nat. 5:136-142.—Murray, Sequoyah, Osage, and Payne counties. *Blarina brevicauda hulophaga* is a synonym of *B. b. carolinensis* (=*B. hylophaga*). Systematics, Distribution, Populations, Morphology. Soricidae.

647 **Jones, J. K., Jr., and J. A. Homan.** 1976. Contribution to a bibliography of Recent Texas mammals, 1971-1980. Occas. Pap. Mus. Tex. Tech Univ. 41:1-21.—Non-annotated bibliography is organized taxonomically. Bibliographies. Mammalia.

648 **Jones, J. K., Jr., and B. Mursaloglu.** 1961. Geographic variation in the harvest mouse, *Reithrodontomys megalotis,* on the central Great Plains and in adjacent regions. Univ. Kans. Publ. Mus. Nat. Hist. 14:9-27.—Distribution of *Reithrodontomys megalotis aztecus* includes Oklahoma Panhandle. Systematics, Distribution, Natural history, Morphology. Cricetidae.

649 **Jones, J. K., Jr., C. J. Young, and D. J. Schmidly.** 1985. Contribution to a bibliography of Recent Texas mammals, 1971-1980. Occas. Pap. Mus. Tex. Tech Univ., 95:1-44.—Update of bibliography concerning mammals in adjacent state to the south. Arranged taxonomically. Bibliographies. Mammalia.

Jones, J. K., Jr., *see* 17, 18, 84, 85, 226, 310, 474, 586, 1151, 1152.

Jordan, H. E., *see* 223, 224.

Joslin, P. W., *see* 880.

650 **Keim, D. B. R.** 1891. Sheridan's troopers on the borders. David McKay, Philadelphia. 308 pp.—*Cervus elaphus indigenous* in Wichita Mountains apparently extirpated around 1881. History, Natural history. Cervidae.

Kellogg, R., *see* 802.

Kellogg, W. H., *see* 851.

Kelson, K. R., *see* 478, 479.

651 **Kennedy, M. L.** 1975. Geographic variation in Ord's kangaroo rat, *Dipodomys ordii.* Ph.D. Dissertation, Univ. Oklahoma, Norman. 120 pp.—Analysis of *Dipodomys ordii* includes Oklahoma specimens. Systematics, Ecology, Morphology. Heteromyidae.

652 **Kennedy, M. L., M. L. Beck, and T. L. Best.** 1980. Intraspecific morphologic variation in Ord's kangaroo rat, *Dipodomys ordii,* from Oklahoma. J. Mammal. 61:311-319.—Variation between central and western Oklahoma populations of *Dipodomys ordii* is clinal. Systematics, Distribution, Morphology. Heteromyidae.

653 **Kennedy, M. L., and G. A. Engbretson.** 1974. Flight speeds of the cave bat. Proc. Okla. Acad. Sci. 54:12-13.—*Myotis velifer* in Harmon County. Flight speed estimates of 10.77 mph and 11.04 mph. Natural history. Vespertilionidae.

654 **Kennedy, M. L., P. L. Leberg, and G. D. Baumgardner.** 1986. Morphologic variation in the coyote, *Canis latrans,* in the southern United States. Southwest. Nat. 31:139-148.—Morphometric study of sexual dimorphism and geographic variation of coyotes. Specimens included from Oklahoma. Systematics, Distribution, Morphology. Canidae.

655 **Kennedy, M. L., and S. L. Lindsay.** 1984. Morphologic variation in the raccoon, *Procyon lotor,* and its relationship to genic and environmental variation. J. Mammal. 65:195-205.—Assessment of sexual and

geographic variation in *Procyon lotor* in Oklahoma and southeastern United States. Systematics, Genetics, Morphology. Procyonidae.

656 **Kennedy, M. L., and G. D. Schnell.** 1978. Geographic variation and sexual dimorphism in Ord's kangaroo rat, *Dipodomys ordii.* J. Mammal. 59:45-59.—*Dipodomys ordii.* Systematics, Distribution, Morphology. Heteromyidae.

Kennedy, M. L., *see* 120, 131, 240, 891, 903.

657 **Kerr, H.** 1964. The Bartow mammoth site. Okla. Anthropol. Soc. Newsl. 12(5):4-8.—Mammoth collected in Woodward County. Paleontology. Elephantidae.

658 **Kessel, E. L.** 1952. New host records for *Trichobius corynorhini* (Diptera: Streblidae). Wasmann J. Biol. 10:7-8.—Bat fly (*Trichobius corynorhini*) collected from *Corynorhinus rafinesquii* (=*Plecotus townsendii*) *pallescens* and *Tadarida mexicana* (=*brasiliensis*) in Woods County. Distribution, Parasites. Vespertilionidae, Molossidae.

Keyes, N., *see* 307.

659 **Kilgore, D. L., Jr.** 1969. An ecological study of the swift fox (*Vulpes velox*) in the Oklahoma Panhandle. Am. Midl. Nat. 81:512-534.—Study in Beaver County includes *Vulpes velox, Peromyscus maniculatus, Mephitis mephitis, Cryptotis parva, Sylvilagus* spp., *Lepus californicus, Cynomys ludovicianus, Spermophilus tridecemlineatus, Perognathus flavus, Perognathus* (=*Chaetodipus*) *hispidus, Reithrodontomys megalotis, Sigmodon hispidus,* and *Mus musculus.* History, Distribution, Natural history, Ecology, Populations, Parasites, Morphology. Canidae, Mammalia.

Kilgore, D. L., Jr., *see* 921.

660 **Kilpatrick, C. W.** 1984. Molecular evolution of the Texas mouse, *Peromyscus attwateri.* Festschrift for Walter W. Dalquest in honor of his sixty-sixth birthday (N. V. Horner, ed.). Biology Department, Midwestern State Univ., Wichita Falls, Texas. 163 pp.—Study includes specimens of *Peromyscus attwateri* from Comanche County and *P. boylii* from Cimarron County. Systematics, Genetics. Cricetidae.

661 **Kilpatrick, C. W., and W. Caire.** 1973. The first record of the encinal mouse, *Peromyscus pectoralis,* for Oklahoma, and additional records for northcentral Texas. Southwest. Nat. 18:351.—*Peromyscus pectoralis, P. boylii, P. attwateri, P. leucopus,* and *Neotoma floridana.* Love County, Oklahoma and three Texas counties. Distribution, Ecology, Morphology. Cricetidae.

662 **Kilpatrick, C. W., and E. G. Zimmerman.** 1975. Genetic variation and systematics of four species of mice of the *Peromyscus boylii* species group. Syst. Zool. 24:143-162.—*Peromyscus attwateri, P. boylii, P. pectoralis* and *P. polius.* Cimarron, Comanche and Love counties, Oklahoma, as well as Texas, New Mexico, Utah, Kansas, and Mexico. Considerable chromosomal variation occurred among populations of *P. boylii. Peromyscus attwateri* probably arose from *P. boylii.* Affinities of *P. pectoralis* probably are not with *boylii* group. Systematics, Genetics. Cricetidae.

663 **Kilpatrick, C. W., and E. G. Zimmerman.** 1976. Biochemical variation and systematics of *Peromyscus pectoralis.* J. Mammal. 57:506-522. —Electrophoretic study of *Peromyscus pectoralis* from Love County, Oklahoma, as well as Texas and Mexico. Systematics, Populations, Genetics. Cricetidae.

664 **Kilpatrick, C. W., and E. G. Zimmerman.** 1976. Hemoglobin polymorphism in the encinal mouse, *Peromyscus pectoralis.* Biochem. Genet. 14:137-143.—Populations of *Peromyscus pectoralis* from Love County were monomorphic for hemoglobin type. Systematics, Genetics. Cricetidae.

Kilpatrick, C. W., *see* 1204, 1205.

665 **Kimsey, A.** 1953. Nutria. Okla. Wildl. News 9(3):10-11.—*Myocastor coypus* raised commercially in Caddo and Comanche counties. Ecology, Wildlife management. Myocastoridae.

666 **King, J. A., ed.** 1968. Biology of *Peromyscus* (Rodentia). Spec. Publ. Am. Soc. Mammal. 2:1-593.—Systematics, Paleontology, History, Distribution, Natural history, Behavior, Ecology, Populations, Parasites, Genetics, Morphology, Physiology, Bibliographies. Cricetidae.

667 **King, O. M.** 1956. A list of mammals collected in west central Oklahoma. Proc. Okla. Acad. Sci. 37:52-53.—Annotated list of 33 species from Custer, Alfalfa, Blaine, Caddo, Kiowa, Major, and Washita counties. Distribution. Mammalia.

Kinman, K. E., *see* 600.

668 **Kirkpatrick, R. D.** 1964. Evaluation of the lens weight as an aging technique for the cotton rat, *Sigmodon hispidus* Say and Ord. Ph.D. Dissertation, Oklahoma State Univ., Stillwater. 43 pp.—Payne County study of *Sigmodon hispidus* indicates that lens weight may be an accurate age indicator under only limited conditions. Morphology, Techniques. Cricetidae.

669 **Kirkpatrick, R. D.** 1965. Litter size and fetus numbers in the cotton rat. J. Mammal. 46:514.—Hispid cotton rats (*Sigmodon hispidus*) raised in enclosures. Adults obtained in Payne County. Natural history. Cricetidae.

670 **Kitts, D. B.** 1957. A Pliocene vertebrate fauna from Ellis County, Oklahoma. Okla. Geol. Surv. Circ. 45:1-27.—Ellis County site yielded remains of *Osteoborus, Aelurodon, Albanosmilus, Serridentinus, Pliohippus, Neohipparion, Aphelops,* and *Prostennhops.* Systematics, Paleontology. Canidae, Felidae, Gomphotheriidae, Equidae, Rhinocerotidae.

671 **Kitts, D. B.** 1958. *Nimravides,* a new genus of Felidae from the Pliocene of California, Texas and Oklahoma. J. Mammal. 39:368-375.—Distribution of *Nimravides thinobates* includes Ellis County site. Systematics, Paleontology. Felidae.

672 **Kitts, D. B.** 1958. A saber-toothed cat, *Smilodon californicus* Bovard, from Logan County, Oklahoma. Okla. Geol. Notes 18:19-23.—*Smilodon californicus* discovered in Logan County, with some discussion of relationships of North American fossil cats. Systematics, Paleontology. Felidae.

673 **Kitts, D. B.** 1964. *Aelurodon,* an addition to the Durham local fauna, Roger Mills County, Oklahoma. Okla. Geol. Notes 24:76-78.—*Aelurodon* added to the previously described mammalian fauna of site in Roger Mills County. Paleontology. Canidae.

674 **Kitts, D. B., and C. C. Black.** 1959. A Pliocene vertebrate local fauna from Roger Mills County, Oklahoma. Okla. Geol. Surv. Circ. 48:27-47.—Roger Mills County site. Remains of rodents, carnivores, horses, and camels. Paleontology. Mylagaulidae, Heteromyidae, Canidae, Mustelidae, Equidae, Camelidae.

675 **Kitts, D. B., and A. J. Myers.** 1959. A Pliocene badger *Pliotaxidea nevadensis* (Butterworth) from Harper County, Oklahoma. Okla. Geol. Notes 19:143-146.—Harper County site yielded remains of *Pliotaxidea nevadensis, Nannippus, Neohipparion,* and *Pliohippus.* Discussion of relationships of North American fossil badgers. Systematics, Paleontology. Mustelidae, Equidae.

676 **Klingener, D.** 1963. Dental evolution in *Zapus.* J. Mammal. 44:248-260.—Fossil jumping mice (*Zapus*) from deposits in Kansas and Oklahoma. Systematics, Paleontology, Morphology. Zapodidae.

677 **Klippel, W. E., and P. W. Parmalee.** 1984. Armadillos in North American Late Pleistocene contexts. Pages 149-160 *in* Contributions

in Quaternary vertebrate paleontology: A volume in memorial to John E. Guilday (H. H. Genoways and M. R. Dawson, eds.). Carnegie Mus. Nat. Hist. Spec. Publ. No. 8. 538 pp.—Includes *Dasypus bellus* from Bar M locality in Harper County. Systematics, Paleontology. Dasypodidae.

Knopf, F. L., *see* 842.

678 **Knox, M. V. B.** 1875. Kansas Mammalia. Trans. Kans. Acad. Sci. 4:18-22.—Some comments on areas adjacent to Oklahoma. Distribution and status of 61 species known to occur in Kansas. Distribution. Mammalia.

679 **Kocan, A. A.** 1983. White-tailed deer herd health evaluation. Okla. Dep. Wildl. Cons. Job 01, Study 02, W-130-R. 145 pp.—For *Odocoileus virginianus.* Parasites, Wildlife management. Cervidae.

680 **Kocan, A. A., E. F. Blouin, and B. L. Glenn.** 1985. Hematologic and serum chemical values for free-ranging bobcats, *Felis rufus* (Schreber), with reference to animals with natural infections of *Cytauxzoon felis* Kier, 1979. J. Wildl. Dis. 21:190-192.—*Felis rufus.* Parasites. Felidae.

681 **Kocan, A. A., B. L. Glenn, T. R. Thedford, R. Doyle, K. Waldrup, G. Kubat, and M. G. Shaw.** 1981. Effects of chemical immobilization on hematologic and serum chemical values in captive white-tailed deer. J. Am. Vet. Med. Assoc. 179:1153-1156.—Physiology, Techniques, Wildlife management. Cervidae.

682 **Kocan, A. A., M. G. Shaw, W. C. Edwards, and J. H. Eve.** 1980. Heavy metal concentrations in the kidneys of white-tailed deer in Oklahoma. J. Wildl. Dis. 16:593-596.—Renal concentrations of lead, arsenic, mercury and cadmium were measured in *Odocoileus virginianus* from 10 counties. Ecology, Physiology, Economics, Wildlife management. Cervidae.

683 **Kocan, A. A., M. G. Shaw, K. S. Waldrup, and G. J. Kubat.** 1982. Distribution of *Parelaphostrongylus tenuis* (Nematoda: Metastrongyloidea) in white-tailed deer from Oklahoma. J. Wildl. Dis. 18:457-460.—Meningeal worms found in *Odocoileus virginianus* only in eastern forest areas of Oklahoma. Parasites, Wildlife management. Cervidae.

Kocan, A. A., *see* 427, 998, 1160, 1183.

684 **Koch, H. G.** 1981. Suitability of birds and mammals as hosts for immature stages of the lone star tick, *Amblyomma americanum* (Acari: Ixodidae). J. Med. Entomol. 18:93-98.—Feeding studies of lone star

ticks from LeFlore County were conducted on *Didelphis virginiana, Marmota monax, Sciurus niger, Peromyscus maniculatus, Neotoma floridana, Mephitis mephitis,* and *Procyon lotor.* Parasites, Economics. Didelphidae, Sciuridae, Cricetidae, Mustelidae, Procyonidae.

685 **Koch, H. G., and J. C. Dunn.** 1980. Ticks collected from small and medium-sized wildlife in LeFlore County, Oklahoma. Southwest. Entomol. 5:214-221.—A total of 32 of 67 species of birds and mammals sampled from April to September 1978 and 1979 were positive for ticks. About 75% of all ticks collected were lone star ticks (*Amblyomma americanum*). All stages of this tick common on coyotes (*Canis latrans*). Parasites. Canidae, Mammalia.

Koeppl, J. W., *see* 600.

686 **Koford, C. B.** 1958. Prairie dogs, whitefaces and blue grama. Wildl. Monogr. 3:1-78.—Extensive study of biology of *Cynomys ludovicianus,* with some observations from Comanche County. History, Distribution, Natural history, Behavior, Ecology, Populations, Economics, Wildlife management. Sciuridae.

687 **Kohls, G. M.** 1954. New distributional records for *Trichobius corynorhini* (Diptera: Streblidae). Wasmann J. Biol. 12:27-28.—Bat fly *Trichobius corynorhini* collected from *Corynorhinus rafinesquii* (=*Plecotus townsendii*) *pallescens* in Woodward County. Distribution, Parasites. Vespertilionidae.

Kraay, G. J., *see* 868.

688 **Krefting, L. W., and J. B. Fletcher.** 1941. Notes on the cruising method of censusing white-tailed deer in Oklahoma. J. Wildl. Manage. 5:412-415.—Census method employed in southeastern Oklahoma for *Odocoileus virginianus.* Techniques, Wildlife management. Cervidae.

Kruckenberg, S. M., *see* 394.

689 **Krutzsch, P. H.** 1954. North American jumping mouse (genus *Zapus*). Univ. Kans. Publ. Mus. Nat. Hist. 7:349-372.—Distribution of *Zapus hudsonius pallidus* includes northeastern Oklahoma. Systematics, Paleontology, Distribution, Natural history, Behavior, Ecology, Populations, Morphology. Zapodidae.

Kubat, G., *see* 681, 683.

690 **Kunz, T. H.** 1971. Ecology of the cave bat, *Myotis velifer,* in south-central Kansas and northwestern Oklahoma. Ph.D. Dissertation, Univ. Kansas, Lawrence. 148 pp.—Extensive study of *Myotis velifer*

includes localities in Woods, Woodward, and Major counties. History, Distribution, Natural history, Behavior, Ecology, Populations, Morphology, Physiology, Techniques, Bibliographies. Vespertilionidae.

691 **Kunz, T. H.** 1974. Feeding ecology of a temperate insectivorous bat (*Myotis velifer*). Ecology 55:693-711.—Study of *Myotis velifer* in Woodward County, Oklahoma and several Kansas counties. Emergence behavior and foraging patterns varied with age, sex, reproductive condition, season and colony size. Distribution, Natural history, Behavior, Ecology, Populations. Vespertilionidae.

692 **Kunz, T. H.** 1982. *Lasionycteris noctivagans.* Mamm. Species 172:1-5.—Summary of published information on *Lasionycteris noctivagans.* Systematics, Paleontology, Distribution, Natural history, Behavior, Ecology, Parasites, Genetics, Morphology, Bibliographies. Vespertilionidae.

693 **Kunz, T. H., and R. A. Martin.** 1982. *Plecotus townsendii.* Mamm. Species 175:1-6.—*Plecotus townsendii* distribution includes western and northeastern Oklahoma. Summary of information in literature. Systematics, Paleontology, Distribution, Natural history, Behavior, Ecology, Parasites, Genetics, Morphology, Physiology, Wildlife management, Bibliographies. Vespertilionidae.

Kunz, T. H., *see* 376, 623.

694 **Kurten, B., and E. Anderson.** 1980. Pleistocene mammals of North America. Columbia Univ. Press, New York. 442 pp.—Authors list and characterize Pleistocene faunal assemblages for North America, including some in Oklahoma. Also includes an annotated fossil record for 562 mammal species. Paleontology, Distribution, Bibliographies. Mammalia.

Labadies, B., *see* 1043.

695 **Lackey, J. A., D. G. Huckaby, and B. G. Ormiston.** 1985. *Peromyscus leucopus.* Mamm. Species 247:1-10.—Summary of published information on white-footed mouse, *Peromyscus leucopus.* Systematics, Paleontology, History, Distribution, Natural history, Behavior, Ecology, Populations, Parasites, Genetics, Morphology, Bibliographies. Cricetidae.

Lacy, K. E., *see* 209.

696 **Landreth, H. F.** 1972. Ecology of Norway rats, *Rattus norvegicus,* on a deserted farm in western Oklahoma. Proc. Okla. Acad. Sci. 52:45-48. —*Rattus norvegicus, Sigmodon hispidus, Neotoma floridana, Sylvilagus*

floridanus, Mus musculus, Didelphis virginiana, Perognathus (=*Chaetodipus*) *hispidus, Mephitis mephitis,* and *Peromyscus leucopus* in Custer County. Distribution, Behavior, Ecology, Populations. Didelphidae, Leporidae, Heteromyidae, Cricetidae, Muridae, Mustelidae.

697 **Landreth, H. F., M. T. Christensen, L. J. Bussjaeger, A. J. Stanley, and J. E. Allison.** 1976. Influence of genetically sterile males on fecundity of Norway rats. Biol. Reprod. 15:390-395.—Introduction of genetically sterile male *Rattus norvegicus* into caged populations of wild rats from Oklahoma County resulted in reduced female fecundity. Natural history, Behavior, Populations, Genetics. Muridae.

Landreth, H. F., Jr., *see* 460.

698 **Lantz, D. E.** 1907. Additions and corrections to the list of Kansas mammals. Trans. Kans. Acad. Sci. 20:214-217.—Includes comments on distributions in adjacent counties in Oklahoma. Distribution. Mammalia.

699 **Lantz, G.** 1971. The coyote: Free spirit of the prairie. Outdoor Okla. 27(7):10-11.—*Canis latrans.* Account for the layman on the man-versus-coyote conflict. Natural history, Economics. Canidae.

Lauzon, S. E., *see* 607.

700 **LaVal, R. K.** 1967. Records of bats from the southeastern United States. J. Mammal. 48:645-648.—*Myotis austroriparius* in Pushmataha County. Distribution. Vespertilionidae.

701 **LaVal, R. K.** 1970. Infraspecific relationships of bats of the species *Myotis austroriparius.* J. Mammal. 51:542-552.—*Myotis austroriparius* occurs in LeFlore, McCurtain, and Pushmataha counties. Systematics, Distribution, Morphology. Vespertilionidae.

702 **Lawton, S. P.** 1968. The Duncan-Wilson bluff shelter: A stratified site of the Southern Plains. Bull. Okla. Anthropol. Soc. 16:1-94.—Excavation of Caddo County site produced bones of about 22 mammal species. Paleontology. Mammalia.

Leberg, P. L., *see* 654.

703 **Lechleitner, R. R.** 1969. Wild mammals of Colorado. Their appearance, habits, distribution, and abundance. Pruett Publishing Co., Boulder, Colorado.—Covers animals of adjacent state. Systematics, History, Distribution, Behavior, Ecology, Morphology. Mammalia.

704 **Lee, M. R., D. J. Schmidly, and C. C. Huheey.** 1972. Chromosomal variation in certain populations of *Peromyscus boylii* and its systematic implications. J. Mammal. 53:697-707.—*Peromyscus boylii attwateri* accorded specific status based upon karyotypic differences. Systematics, Genetics, Morphology. Cricetidae.

705 **Leonhardy, F. C.** 1965. New radiocarbon dates for the Domebo site. Okla. Anthropol. Soc. Newsl. 13(6):2.—Age of the Domebo mammoth estimated at 11,200 years. Paleontology. Elephantidae.

706 **Leonhardy, F. C.** 1966. Domebo: A Paleo-Indian mammoth kill in the Prairie Plains. Contr. Mus. Great Plains 1:1-53.—Includes discussion of Caddo County Pleistocene mammal fauna by M. G. Mehl and B. H. Slaughter. Paleontology. Geomyidae, Heteromyidae, Cricetidae, Bovidae, Elephantidae.

707 **Leopold, A. L., L. K. Sowls, and D. L. Spencer.** 1947. A survey of overpopulated deer ranges in the United States. J. Wildl. Manage. 11:162-177.—Incipient irruptions of *Odocoileus virginianus* populations predicted in Wichita Mountains Wildlife Refuge and Lake Murray State Park. Distribution, Ecology, Populations, Wildlife management. Cervidae.

708 **Leppla, R. R., II.** 1980. An assessment of damage to pecans by wildlife in central Oklahoma. M.S. Thesis, Oklahoma State Univ., Stillwater. 71 pp.—Lincoln County study showed *Sciurus niger* to be only significant mammalian predator on pecans. Natural history, Ecology, Wildlife management. Sciuridae, Mammalia.

Levescy, B., *see* 207.

709 **Lewis, J. C.** 1968. Evidence of mountain lions in the Ozark, Boston and Ouachita Mountains. Proc. Okla. Acad. Sci. 49:182-184.—*Felis concolor.* One recent specimen, several observations from eastern Oklahoma. Distribution. Felidae.

710 **Lewis, J. C.** 1968. The lion in seven mile slough. Outdoor Okla. 24(7):16-17.—*Felis concolor* in McIntosh County. Account for the layman. Adult female found dead. Distribution, Ecology, Morphology, Popular. Felidae.

711 **Lewis, J. C.** 1969. Evidence of mountain lions in the Ozarks and adjacent areas, 1948-1968. J. Mammal. 50:371-372.—Reports on 1968 specimen of mountain lion (*Felis concolor*) from McIntosh County and summarizes records for region over previous 20 years. Distribution. Felidae.

712 **Lewis, J. C.** 1972. Factors influencing reports of rabid animals in Oklahoma. J. Wildl. Dis. 8:245-251.—Striped skunk (*Mephitis mephitis*) is primary vector for rabies in Oklahoma, but reports of rabid pets are probably better indicators of rabies problem areas. Parasites, Wildlife management. Mustelidae.

713 **Lewis, J. C.** 1973. Additional records of black-footed ferret in Oklahoma. Southwest. Nat. 18:350.—*Mustela nigripes* and *Cynomys ludovicianus* from Cleveland, Woods, and Texas counties. Distribution, Ecology. Sciuridae, Mustelidae.

714 **Lewis, J. C.** 1974. Project 977: Status survey of the black-footed ferret. World Wildl. Yearb. 1973-1974:243.—Discussion of possible ferret (*Mustela nigripes*) signs in Texas and Cimarron counties. Distribution, Wildlife management. Mustelidae.

715 **Lewis, J. C. (and Rare and Endangered Species of Oklahoma Committee).** 1975. Rare and endangered vertebrates and plants of Oklahoma. Rare and Endangered Species of Oklahoma Committee, Stillwater. 44 pp.—Lists distributions and habitats for 25 species of mammals thought to be endangered or rare in Oklahoma. Distribution, Ecology, Populations, Wildlife management, Popular. Mammalia.

716 **Lewis, J. C., and F. D. Hassien.** 1974. Status of prairie dogs and black-footed ferrets in Oklahoma. Proc. Okla. Acad. Sci. 54:20-24.—*Cynomys ludovicianus, Mustela nigripes, Geomys bursarius, Mustela frenata,* and *Sylvilagus audubonii.* Ferrets extinct or rare in Panhandle. History, Distribution, Ecology, Populations, Economics. Leporidae, Sciuridae, Geomyidae, Mustelidae.

717 **Lewis, J. C., and F. D. Hassien.** 1974. Status of prairie dogs and surveys for black-footed ferrets in Oklahoma. Pages 60-75 *in* Proceedings of the black-footed ferret and prairie dog workshop (R. L. Linder and C. N. Hillman, eds.). South Dakota State Univ., Brookings.—Extensive survey of *Cynomys ludovicianus* towns in Cimarron, Beaver, and Texas counties indicated that *Mustela nigripes* probably extirpated from state. History, Distribution, Populations, Techniques, Wildlife management. Sciuridae, Mustelidae.

718 **Lewis, J. C., E. H. McIlvain, R. McVickers, and B. Peterson.** 1979. Techniques used to establish and limit prairie dog towns. Proc. Okla. Acad. Sci. 59:27-30.—*Cynomys ludovicianus* populations established successfully. Natural history, Populations. Sciuridae.

Lewis, J. C., *see* 111, 112, 203, 383, 842, 907.

Lindsay, S. L., *see* 655.

719 **Lindzey, J. S.** 1951. The white-tailed deer in Oklahoma: Ecology, management and production. Ph.D. Dissertation, Oklahoma A and M Coll., Stillwater. 159 pp.—McCurtain County study of habitat requirements, with management recommendations for *Odocoileus virginianus.* Predation by *Canis familiaris, C. rufus,* and *Lynx rufus* found to be unimportant. History, Distribution, Natural history, Behavior, Ecology, Populations, Parasites, Techniques, Economics, Wildlife management. Canidae, Felidae, Cervidae.

720 **Lindzey, J. S.** 1951. The white-tailed deer in Oklahoma: Management and production. Okla. Dep. Wildl. Cons. W-037-R. 105 pp.—Description of four major ranges of *Odocoileus virginianus* in Oklahoma, with emphasis on McCurtain County. Natural history, Ecology, Populations, Parasites, Economics, Wildlife management. Cervidae.

721 **Lintz, C.** 1976. Test excavations at the Kenton bison kill site (Ci-81), Cimarron County, Oklahoma. Okla. Anthropol. Soc. Newsl. 24(7):2-7.—Test pits dug at site in Cimarron County, where *Bison* bones were found protruding from bluff. Paleontology. Bovidae.

722 **Lintz, C.** 1982. Archeological testing within the Kickapoo Nations watershed, Lincoln County, Oklahoma. Okla. Conservation Commission, Archeol. Research Rep. No. 8. 237 pp.—Small sample of faunal remains. Paleontology. Leporidae, Canidae, Cervidae, Bovidae.

723 **Lintz, C., and S. Hall.** 1983. The geomorphology and archeology of Carnegie Canyon Ft. Cobb laterals watershed, Caddo Co., Oklahoma. Okla. Conservation Commission, Archeol. Research Rep. No. 10. 221 pp.—Five bone concentrations including four of bison (*Bison bison*) and one of white-tailed deer (*Odocoileus virginianus*). Paleontology. Cervidae, Bovidae.

Linzey, A. V., *see* 1185.

724 **Linzey, D. W., and R. L. Packard.** 1977. *Ochrotomys nuttalli.* Mamm. Species 75:1-6.—Distribution of *Ochrotomys nuttalli* includes extreme eastern Oklahoma. Summary of information in literature. Systematics, Paleontology, Distribution, Natural history, Behavior, Ecology, Populations, Parasites, Morphology, Physiology, Bibliographies. Cricetidae.

725 **Litvaitis, J.** 1979. Collared coyotes. Outdoor Okla. 35(9):17-19.—Popular article reporting on radio-tracking studies of *Canis latrans* in

and adjacent to the Wichita Mountains National Wildlife Refuge. Natural history, Behavior, Ecology, Techniques, Popular. Canidae.

726 **Litvaitis, J. A.** 1978. Movements and habitat use of coyotes on the Wichita Mountains National Wildlife Refuge. M.S. Thesis, Oklahoma State Univ., Stillwater. 70 pp.—Comanche County study assessed home-range size and utilization patterns of *Canis latrans.* Rodents and rabbits are primary prey items. History, Natural history, Behavior, Ecology, Wildlife management. Leporidae, Cricetidae, Canidae, Cervidae.

727 **Litvaitis, J. A.** 1981. A comparison of coyote and bobcat food habits in the Wichita Mountains, Oklahoma. Proc. Okla. Acad. Sci. 61:81-82.—Rodents were most common foods of both *Canis latrans* and *Felis rufus* in Comanche County. Natural history, Ecology. Canidae, Felidae, Mammalia.

728 **Litvaitis, J. A., and W. S. Bartush.** 1980. Coyote-deer interactions during the fawning season in Oklahoma. Southwest. Nat. 25:117-118. —*Odocoileus virginianus* doe pursued two *Canis latrans* in fawning season. Behavior. Canidae, Cervidae.

729 **Litvaitis, J. A., and J. H. Shaw.** 1980. Coyote movements, habitat use, and food habits in southwestern Oklahoma. J. Wildl. Manage. 44:62-68.—*Canis latrans* daily and seasonal movements discussed. Coyotes preferentially used savannah over prairie. Main food items were rodents. Natural history, Behavior, Ecology. Dasypodidae, Leporidae, Cricetidae, Canidae, Cervidae.

Lodes, M. J., *see* 1110.

730 **Logan, T. H.** 1970. Study of Rio Grande wild turkey by radio telemetry. Okla. Dep. Wildl. Cons. Fin. W-086-R-04. 46 pp.—Study in Roger Mills and Beckham counties indicates predation on turkeys by *Lynx* (=*Felis*) *rufus.* Ecology. Felidae.

731 **Logan, T. H.** 1973. Study of white-tailed deer fawn mortality on Cookson Hills Deer Refuge, eastern Oklahoma. Proc. Annu. Conf. Southeast. Assoc. Game Fish Comm. 26:27-39.—Study of mortality factors in *Odocoileus virginianus* populations. Natural history, Populations, Parasites. Cervidae.

732 **Logan, T. H.** 1976. Doe harvest: A biological necessity. Outdoor Okla. 32(11):8-10.—*Odocoileus virginianus.* Account for the layman. Populations, Parasites, Wildlife management, Popular. Cervidae.

733 **Long, C. A.** 1961. Woodchuck in Oklahoma and southeastern Kansas. J. Mammal. 42:255-256.—*Marmota monax* in Ottawa County. Distribution. Sciuridae.

734 **Long, C. A.** 1972. Taxonomic revision of the North American badger, *Taxidea taxus.* J. Mammal. 53:725-759.—*Taxidea taxus* specimens at U.S. National Museum examined, including specimens from Texas and Ellis counties. Systematics, Distribution. Mustelidae.

735 **Long, C. A.** 1973. *Taxidea taxus.* Mamm. Species 26:1-4.—*Taxidea taxus* distributed throughout Oklahoma. Summary of information in literature. Systematics, Paleontology, History, Distribution, Natural history, Behavior, Ecology, Parasites, Morphology, Physiology, Bibliographies. Mustelidae.

736 **Long, C. A.** 1975. Molt in the North American badger, *Taxidea taxus.* J. Mammal. 56:921-924.—*Taxidea taxus.* Detailed pelage descriptions. Specimens from western United States, including Oklahoma. Natural history, Morphology. Mustelidae.

737 **Long, C. A., and C. F. Long.** 1964. Range extension of the gray fox, *Urocyon cinereoargenteus.* Southwest. Nat. 9:108-109.—*Urocyon cinereoargenteus ocythous* in McCurtain County. Distribution. Canidae.

Long, C. F., *see* 737.

738 **Looney, M. W.** 1972. Predation on bats by hawks and owls. Bull. Okla. Ornithol. Soc. 5:1-4.—*Tadarida brasiliensis* in Major County preyed upon by barn owls (*Tyto alba*). Other accounts of predation on bats by various birds of prey (*Falco peregrinus, F. sparverius, Bubo virginianus, Buteo jamaicensis, B. lagopus,* and *Circus hudsonius*). Ecology. Molossidae.

739 **Lopez, D. R.** 1973. The Wybark site. A Late Prehistoric village complex in the Three Forks locale, Muskogee County, Oklahoma. Bull. Okla. Anthropol. Soc. 22:11-126.—Bones of 13 mammal species recovered. Most common were bison (*Bison bison*) and white-tailed deer (*Odocoileus virginianus*). Paleontology. Cervidae, Bovidae, Mammalia.

740 **Loring, J. A.** 1905. The Wichita buffalo range. Annu. Rep. New York Zool. Soc. 10:181-200.—Survey of Wichita Game Reserve for use as refuge for *Bison bison, Antilocapra americana, Cervus elaphus, Odocoileus virginianus,* and *O. hemionus.* Distribution, History, Wildlife management. Canidae, Procyonidae, Mustelidae, Cervidae, Antilocapridae, Bovidae.

741 **Lotze, J.-H., and S. Anderson.** 1979. *Procyon lotor.* Mamm. Species 119:1-8.—*Procyon lotor* distribution includes all of Oklahoma. Summary of information in literature. Systematics, Paleontology, Distribution, Natural history, Behavior, Ecology, Populations, Parasites, Genetics, Morphology, Physiology, Economics, Bibliographies. Procyonidae.

Loughry, W. J., *see* 854.

742 **Lowrey, C.** 1971. Elk status report. Outdoor Okla. 27(10):10-12.—*Cervus elaphus.* Account for the layman. Distribution, Wildlife management, Popular. Cervidae.

Lowry, F., *see* 220, 1118, 289, 1046.

743 **MacClintock, D.** 1970. Squirrels of North America. Van Nostrand Reinhold Co., New York.—Distribution, Ecology, Behavior. Sciuridae.

744 **MacFadden, B. J.** 1984. Systematics and phylogeny of *Hipparion, Neohipparion, Nannippus,* and *Cormohipparion* (Mammalia, Equidae) from the Miocene and Pliocene of the New World. Bull. Am. Mus. Nat. Hist. 179:1-195.—Included specimens from Ellis and Texas counties in Oklahoma. Systematics, Paleontology, Distribution, Morphology. Equidae.

745 **Macy, R. W.** 1933. A review of the trematode family Urotrematidae with the description of a new genus and two new species. Trans. Am. Microsc. Soc. 52:247-254.—Named trematode from small intestine of Brazilian free-tailed bat (*Tadarida brasiliensis*) taken in Freedom, Oklahoma. Parasites. Molossidae.

Mahaffey, J., *see* 640.

746 **Marcy, R. B.** 1854. Exploration of the Red River of Louisiana in the year 1852 (Appendix F. Zoology. Pages 200-201, Mammals.). B. Tucker, Senate Printer, Washington. 310 pp.—List of 24 species that occur in present Oklahoma. History, Distribution. Mammalia.

Marler, R. J., *see* 394.

747 **Martin, A. C., H. S. Zim, and A. L. Nelson.** 1951. American wildlife and plants, a guide to wildlife food habits. McGraw-Hill Book Co., New York. 500 pp.—Indicates that bison (*Bison bison*) in Oklahoma consumed mostly blue gramma, sand dropseed, little bluestem, and windmill grass. Natural history, Behavior. Bovidae, Mammalia.

748 **Martin, C. O., and D. J. Schmidly.** 1982. Taxonomic review of the pallid bat, *Antrozous pallidus* (Le Conte). Spec. Publ. Mus. Tex. Tech Univ. 18:1-48.—Analysis of geographic and nongeographic variation of pallid bats (*Antrozous pallidus*) using skin and skull characters. Some Oklahoma specimens included. Systematics, Distribution, Morphology. Vespertilionidae.

749 **Martin, L. D.** 1975. Scimitar-toothed cats, *Machairodus* and *Nimravides,* from the Pliocene of Kansas and Nebraska. Bull. Univ. Nebr. State Mus. 10:55-63.—Comments on relationships of *Nimravides thinobates* to *Machairodus coloradensis.* Oklahoma specimens considered. Systematics, Paleontology. Felidae.

Martin, L. D., *see* 963, 964.

Martin, R. A., *see* 693.

750 **Martin, R. E., and K. G. Matocha.** 1972. Distributional status of the kangaroo rat, *Dipodomys elator.* J. Mammal. 53:873-877.—Specimens of *Dipodomys elator* taken from Hardeman County, Texas. Status in Oklahoma discussed. Distribution. Heteromyidae.

751 **Martin, R. E., and J. R. Preston.** 1968. The mammals of Harmon County, Oklahoma. Proc. Okla. Acad. Sci. 49:42-60.—Forty-one species listed. Thirty-eight substantiated as present in Harmon County. History, Distribution, Ecology. Mammalia.

Martin, R. E., *see* 890.

752 **Martin, R. P.** 1965. Effects of herbicide applications on wildlife populations. M.S. Thesis, Oklahoma State Univ., Stillwater. 33 pp.—Study in Lincoln and Creek counties indicates that *Sylvilagus floridanus* may not be affected by application of herbicide to forest areas, but *Sciurus niger* populations decline markedly. Ecology, Populations, Wildlife management. Leporidae, Sciuridae.

Martin, R. P., *see* 138, 225.

Mascarello, J. T., *see* 99.

Masters, R., *see* 991, 992, 1119, 1120.

753 **Matocha, K. G.** 1984. Variation of the trill calls of three species of *Spermophilus.* Pages 215-223 *in* Contributions in mammalogy in honor of Robert L. Packard (R. E. Martin and B. R. Chapman, eds.). Spec. Publ. Mus. Tex. Tech Univ. 22:1-234.—Intraspecific comparisons of vocalizations of thirteen-lined ground squirrels (*Spermophilus*

tridecemlineatus) involved two Oklahoma localities (Guymon and Woodward). Behavior. Sciuridae.

Matocha, K. G., *see* 750.

Maughan, O. E., *see* 145.

May, W., *see* 913.

754 **Mayo, M. B.** 1982. Archaeological investigations at the Non Caves (34Hu-59), Bohannon (34Hu-61), and Ashland (34Hu-62) sites in Hughes County, Oklahoma. Archaeol. Research and Management Center, Univ. Oklahoma. Project Rep. Ser. No. 7. 113 pp.—Bone debris included white-tailed deer (*Odocoileus virginianus*) and eastern cottontails (*Sylvilagus floridanus*), as well as a rodent and small mammals that were not identified. Paleontology. Leporidae, Cricetidae, Cervidae.

755 **Mayo, M. B.** 1984. Archeological investigations at the Spring Creek site (34Ln-76), Lincoln County, Oklahoma. Okla. Conservation Commission, Archeol. Research Rep. No. 11. 126 pp.—Numerous remains identified including those of at least seven mammal species. Paleontology. Talpidae, Leporidae, Geomyidae, Cricetidae, Canidae, Cervidae.

756 **McBee, K., and R. J. Baker.** 1982. *Dasypus novemcinctus.* Mamm. Species 162:1-9.—Summary of published information on *Dasypus novemcinctus.* Systematics, Paleontology, History, Distribution, Natural history, Behavior, Ecology, Populations, Parasites, Genetics, Morphology, Bibliographies. Dasypodidae.

757 **McCarley, H.** 1951. Color mutations in a small, partially isolated population of pocket gophers (*Geomys breviceps*). J. Mammal. 32:338-341.—*Geomys breviceps* from Bryan County population exhibits four color mutations. Natural history. Geomyidae.

758 **McCarley, H.** 1952. The ecological relationships of the mammals of Bryan County, Oklahoma. Tex. J. Sci. 4:102-112.—Annotated list of 31 species. Ecological associations. Ecotone between deciduous forest and grasslands. Bryan County is more eastern in its faunal affinities than it is western. Distribution, Ecology. Mammalia.

759 **McCarley, H.** 1954. Natural hybridization in the *Peromyscus leucopus* species group of mice. Evolution 8:314-323.—In areas of sympatry of *Peromyscus leucopus* and *P. gossypinus,* about 2% of specimens apparently were hybrids. Study includes mice from Bryan County. Sys-

tematics, Distribution, Natural history, Genetics, Morphology. Cricetidae.

760 **McCarley, H.** 1960. The rice rat in southern Oklahoma. J. Mammal. 41:130-131.—*Oryzomys palustris texensis* in Marshall County. Distribution, Ecology. Cricetidae.

761 **McCarley, H.** 1961. New locality records for some Oklahoma mammals. Southwest. Nat. 6:108-109.—Accounts of *Blarina brevicauda* (=*hylophaga*), *Tamias striatus venustus, Cynomys ludovicianus, Oryzomys palustris texensis, Ochrotomys nuttalli flammeus, Ondatra zibethicus cinnamominus,* and *Ursus americanus.* Discussions include reference to Marshall, Bryan, Pushmataha, Latimer, McCurtain, Atoka, Carter, Jefferson, Pittsburg, Sequoyah, Cleveland, and LeFlore counties. Distribution, Ecology. Soricidae, Sciuridae, Cricetidae, Ursidae.

762 **McCarley, H.** 1962. The taxonomic status of wild *Canis* (Canidae) in the south central United States. Southwest. Nat. 7:227-235.—*Canis latrans* and *C. niger.* Choctaw, McCurtain, Pushmataha, Latimer, Bryan, Jefferson, Okfuskee, and Atoka counties, Oklahoma, as well as Arkansas, Louisiana, and Texas. *Canis niger* is extant only in Louisiana. *Canis niger rufus* is probably a hybrid (*C. latrans* and *C. n. gregoryi*). Systematics, Distribution, Natural history, Ecology, Populations. Canidae.

763 **McCarley, H.** 1975. Long-distance vocalizations of coyotes (*Canis latrans*). J. Mammal. 56:847-856.—Evoked and spontaneous vocalizations of *Canis latrans* in Texas and Oklahoma. Sonograms. Behavior. Canidae.

764 **McCarley, H., and C. Carley.** 1976. *Canis latrans* and *C. rufus* vocalization: A continuum. Southwest. Nat. 21:399-400.—Sonogram analysis of *Canis latrans* and *C. rufus.* Systematics, Behavior. Canidae.

765 **McCarley, H., and P. Free.** 1962. A new record of the woodchuck from Oklahoma. J. Mammal. 43:271-272.—*Marmota monax* in Latimer County. Distribution. Sciuridae.

McCarley, H., *see* 1075.

766 **McCarty, R.** 1978. *Onychomys leucogaster.* Mamm. Species 87:1-6.—*Onychomys leucogaster* distribution includes western Oklahoma. Summary of information in literature. Systematics, Paleontology, Distribution, Natural history, Behavior, Ecology, Populations, Parasites, Genetics, Morphology, Physiology, Bibliographies. Cricetidae.

767 **McCoy, C. J.** 1960. Albinism in *Tadarida*. J. Mammal. 41:119.—White-haired and albino *Tadarida brasiliensis mexicana* in Woodward, Major, and Woods counties. Natural history. Molossidae.

McCoy, C. J., *see* 1002.

768 **McCulloch, C. Y., Jr.** 1958. Effects of rodents and rabbits on estimates of forage disappearance. Proc. Okla. Acad. Sci. 39:202-204.—*Dipodomys ordii richardsoni, Onychomys leucogaster breviauritus, Peromyscus* spp., *Sigmodon hispidus texianus,* and *Citellus* (=*Spermophilus*) *spilosoma marginatus* in Woodward County. Inventory cages were not effective barriers. Ecology, Wildlife management. Leporidae, Sciuridae, Heteromyidae, Cricetidae.

769 **McCulloch, C. Y., Jr.** 1959. Populations and range effects of rodents on sand sagebrush grasslands of western Oklahoma. Ph.D. Dissertation, Oklahoma State Univ., Stillwater. 159 pp.—Study in Harper and Woodward counties of small mammal communities. Natural history, Behavior, Ecology, Populations, Economics. Sciuridae, Geomyidae, Heteromyidae, Cricetidae.

770 **McCulloch, C. Y., Jr.** 1961. Age classification and weight variations of Ord's kangaroo rat on the southern Great Plains. Southwest. Nat. 6:149-155.—*Dipodomys ordi* (=*ordii*) *richardsoni* in Harper and Woodward counties. Distribution, Natural history, Ecology, Populations, Morphology. Heteromyidae.

771 **McCulloch, C. Y., Jr., and J. M. Inglis.** 1961. Breeding periods of the Ord kangaroo rat. J. Mammal. 42:337-344.—*Dipodomys ordii richardsoni* in Harper and Woodward counties bred mostly in fall and winter. *Sigmodon hispidus* mentioned. Natural history, Populations. Heteromyidae, Cricetidae.

McDaniel, J. S., *see* 263.

772 **McDonald, J. N.** 1981. North American bison. Their classification and evolution. Univ. California Press, Berkeley. 316 pp.—Includes analyses with some specimens from Oklahoma. Systematics, Paleontology, History, Distribution, Natural history, Populations, Morphology, Economics. Bovidae.

773 **McGee, L. E.** 1965. Extension of the range eastward for *Taxidea taxus* (Carnivora: Mustelidae). Southwest. Nat. 10:78.—Badger in Jefferson County. New county record for Oklahoma. Distribution. Mustelidae.

McGuire, S., *see* 213.

McIlvain, E. H., *see* 718.

774 **McKenna, T. M.** 1979. Sodium metabolism during acclimation to water restriction by wild mice, *Mus musculus.* Ph.D. Dissertation, Univ. Oklahoma, Norman. 47 pp.—Study of *Mus musculus* from Cleveland County. Physiology. Muridae.

775 **McKenna, T. M., and H. Haines.** 1981. Sodium metabolism during acclimation to water restriction by wild mice, *Mus musculus.* Am. J. Physiol. 240:R319-R326.—Studies done on stock of *Mus musculus* that originated from animals caught near Norman, Oklahoma. Physiology. Muridae.

McKenna, T. M., *see* 208, 462, 960.

776 **McKinney, T. D., and M. R. Dunbar.** 1976. Weight of adrenal glands in the bobcat (*Lynx rufus*). J. Mammal. 57:378-380.—*Lynx* (=*Felis*) *rufus* specimens from central Oklahoma. Correlations between endocrine function and adrenal weight. Physiology. Felidae.

McKnight, T. J., *see* 974.

777 **McManus, J. J.** 1974. *Didelphis virginiana.* Mamm. Species 40:1-6.—Distribution of *Didelphis virginiana* includes Oklahoma. Summary of information in literature. Systematics, Paleontology, History, Distribution, Natural history, Behavior, Ecology, Populations, Parasites, Genetics, Morphology, Physiology, Economics, Bibliographies. Didelphidae.

778 **McMullen, J. L.** 1978. Mammals of the Duck Creek local fauna, Late Pleistocene of Kansas. J. Mammal. 59:374-386.—Includes listings of mammals recovered from an Oklahoma locality. Paleontology, Distribution, Ecology. Mammalia.

779 **McMurray, F. B.** 1944. Porcupine records in Oklahoma. J. Mammal. 25:413.—*Erethizon epixanthum* (=*dorsatum*) in Comanche, Harper, and Jackson counties. Distribution. Erethizontidae.

780 **McMurray, F. B.** 1945. Three shrews, *Cryptotis parva,* eaten by a feral house cat. J. Mammal. 26:94.—*Felis catus* ate *Cryptotis parva, Peromyscus maniculatus,* and *Reithrodontomys fulvescens.* Ecology. Soricidae, Cricetidae, Felidae.

781 **McMurray, F. B.** 1947. An unusual winter record of *Citellus spilosoma* in Oklahoma. J. Mammal. 28:292.—*Citellus* (=*Spermophilus*) *spilosoma major.* Distribution, Natural history. Sciuridae.

782 **McMurray, F. B., and C. C. Sperry.** 1941. Food of feral house cats in Oklahoma, a progress report. J. Mammal. 22:185-190.—*Felis catus*

ate mostly mammals and a few birds. Ecology, Parasites. Felidae, Mammalia.

McVickers, R., *see* 718.

783 **Mead, J. R.** 1899. *Felis concolor.* Trans. Kans. Acad. Sci. 16:278-279.—Makes comment that mountain lion common along southern border of Kansas, "yet more common in the Indian Territory, now known as Oklahoma." History, Distribution, Natural history. Felidae.

784 **Meade, G. E.** 1953. An early Pleistocene vertebrate fauna from Frederick, Oklahoma. J. Geol. 61:452-460.—Lists and discusses fossil mammals recovered from Holloman gravel pit, Tillman County. Systematics, Paleontology. Mammalia.

785 **Meagher, M.** 1986. *Bison bison.* Mamm. Species 266:1-8.—Summary of published information on bison, *Bison bison.* Systematics, Paleontology, History, Distribution, Natural history, Behavior, Ecology, Populations, Parasites, Genetics, Morphology, Economics, Wildlife management, Bibliographies. Bovidae.

786 **Mech, L. D.** 1974. *Canis lupus.* Mamm. Species 37:1-6.—*Canis lupus* formerly distributed throughout Oklahoma. Summary of information in literature. Systematics, Paleontology, History, Distribution, Natural history, Behavior, Ecology, Populations, Parasites, Genetics, Morphology, Physiology, Techniques, Economics, Wildlife management, Bibliographies. Canidae.

787 **Mehl, M. G.** 1966. The Domebo mammoth: Vertebrate paleomorphology. Contr. Mus. Great Plains 1:27-30.—Discussion of *Mammuthus* specimen from Caddo County. Paleontology. Elephantidae.

Mehlhop, P., *see* 167.

788 **Melchiors, M. A., T. H. Silker, and J. E. Reebs.** 1985. Deer use of young pine plantations in southeastern Oklahoma. J. Wildl. Manage. 49:958-962.—White-tailed deer (*Odocoileus virginianus*) use studied on edge, intermediate, and center zones of large four- and five-year-old pine plantations. Ecology, Wildlife management. Cervidae.

789 **Melchiors, M. A., R. E. Thackston, and D. G. Stobaugh.** 1985. Results of spotlight and helicopter deer surveys. Proc. Annu. Conf. Southeast. Assoc. Fish Wildl. Agencies 39:506-511.—Comparison of methods for evaluating numbers of white-tailed deer (*Odocoileus virginianus*) in young pine plantations in Ouachita Mountains of Oklahoma and Arkansas. Populations, Techniques, Wildlife management. Cervidae.

790 **Melrose, R. R.** 1978. Incised feline jawbones, Le Flore County, Oklahoma. Okla. Anthropol. Soc. Newsl. 26(7):3-4.—Feline jawbones found on shoreline of Wister Lake in LeFlore County. Paleontology. Felidae.

791 **Melton, B. J.** 1922. Killing prairie dogs in Oklahoma. Okla. Agric. Ext. Circ. 162:1-7.—Discussion of various methods of extermination of *Cynomys ludovicianus* in Oklahoma. History, Distribution, Techniques, Economics, Popular. Sciuridae.

792 **Melton, J. E.** 1976. Fluid compartmentation and salt tolerance in two populations of Ord's kangaroo rat (*Dipodomys ordii*). M.S. Thesis, Univ. Oklahoma, Norman. 30 pp.—*Dipodomys ordii oklahomae* and *D. o. richardsoni* from Cleveland and Alfalfa counties, respectively. Both populations have capability to use saline water sources. Ecology, Physiology. Heteromyidae.

Melton, J. E., *see* 462.

793 **Merriam, C. H.** 1892. The geographic distribution of life in North America with special reference to the Mammalia. Proc. Biol. Soc. Wash. 7:1-64.—Includes lists of genera and subgenera of mammals. Oklahoma placed in Sonoran region. Distribution, Ecology. Mammalia.

794 **Merriam, C. H.** 1895. Monographic revision of the pocket gophers. N. Am. Fauna 8:1-258.—*Geomys lutescens* (=*bursarius*) in western Oklahoma, *G. breviceps* (=*bursarius*) in eastern Oklahoma, and *Cratogeomys castanops* in extreme western Panhandle. Systematics, Distribution, Natural history, Behavior, Ecology, Morphology. Geomyidae.

795 **Merriam, C. H.** 1897. Revision of the coyotes or prairie wolves, with descriptions of new forms. Proc. Biol. Soc. Wash. 112:19-33.—Oklahoma coyotes referred to *Canis frustror* (=*latrans*). Systematics, Distribution, Morphology. Canidae.

796 **Merriam, C. H.** 1902. The prairie dog of the Great Plains. U.S. Dep. Agric. Yearb. 1901:257-270.—Distribution of *Cynomys ludovicianus* included western half of Oklahoma. Distribution, Natural history, Behavior, Ecology, Morphology, Techniques, Economics. Sciuridae.

797 **Merrifield, G. C.** 1953. Occurrence of a mountain lion in Oklahoma. Proc. Okla. Acad. Sci. 34:75.—*Felis concolor* near Canton. Plaster cast of track. Distribution, Ecology. Felidae.

Miller, A. F., *see* 514.

798 **Miller, G. S., Jr.** 1907. The families and genera of bats. U.S. Natl. Mus. Bull. 57:1-282.—Systematics, History, Distribution, Morphology. Vespertilionidae, Molossidae, Mammalia.

799 **Miller, G. S., Jr.** 1912. The names of the large wolves of northern and western North America. Smithson. Misc. Collect. 59(15):1-5.—Includes brief mention of *Canis frustror* (=*latrans*). Indicates that type locality is Ft. Gibson, Oklahoma. Systematics. Canidae.

800 **Miller, G. S., Jr.** 1924. List of North American Recent mammals, 1923. U.S. Natl. Mus. Bull. 128:1-673.—Includes list of all mammal type localities from Oklahoma. Systematics, Distribution. Mammalia.

801 **Miller, G. S., Jr., and G. M. Allen.** 1928. The American bats of the genera *Myotis* and *Pizonyx.* U.S. Natl. Mus. Bull. 144:1-218.—Distributions of some *Myotis* species include portions of Oklahoma. Systematics, Distribution. Vespertilionidae.

802 **Miller, G. S., Jr., and R. Kellogg.** 1955. List of North American Recent mammals. U.S. Natl. Mus. Bull. 205:1-954.—Includes list of all mammal type localities in Oklahoma. Systematics, Distribution. Mammalia.

803 **Miller, S. W.** 1931. Notes on some Colorado wood rats. J. Mammal. 12:432.—*Neotoma mexicana* thought to occur in Oklahoma. Distribution. Cricetidae.

804 **Mincolla, J. A.** 1977. The socio-economic impact of the coyote in Oklahoma. M.S. Thesis, Oklahoma State Univ., Stillwater. 119 pp.—*Canis latrans* less preferred by survey respondents than *Procyon lotor, Felis rufus, Sciurus* species, *Odocoileus virginianus, Lepus,* and *Sylvilagus* species. Economics, Wildlife management. Leporidae, Sciuridae, Canidae, Procyonidae, Felidae, Cervidae.

805 **Mitchell, C., and R. D. Burns.** 1963. Mammal records for Oklahoma County. Proc. Okla. Acad. Sci. 44:63-64.—Ten new species records for Oklahoma County. Distribution. Mammalia.

806 **Mogharabi, F., and H. Haines.** 1973. Dehydration and body fluid regulation in the thirteen-lined ground squirrel and laboratory rat. Am. J. Physiol. 224:1218-1222.—*Spermophilus tridecemlineatus* used in study were caught near Norman, Oklahoma. Physiology. Sciuridae, Muridae.

807 **Mohr, C. E.** 1952. A survey of bat banding in North America, 1932-1951. Bull. Natl. Speleol. Soc. 14:3-13.—Distribution, Populations, Techniques. Vespertilionidae, Molossidae.

808 **Mohr, C. E.** 1972. The status of threatened species of cave-dwelling bats. Bull. Natl. Speleol. Soc. 34:33-47.—Includes comments on Oklahoma populations of *Tadarida brasiliensis* and *Myotis grisescens.* Distribution, Ecology, Populations, Parasites, Wildlife management. Vespertilionidae, Molossidae.

Moore, G. A., *see* 164.

Moore, M., *see* 876.

809 **Morgan, M.** 1985. A snowy owl in Comanche County, Oklahoma. Bull. Okla. Ornithol. Soc. 18:29-30.—Remains of hispid cotton rat (*Sigmodon hispidus*) found in pellets of snowy owl (*Nyctea scandiaca*). Distribution, Ecology. Cricetidae.

810 **Morrison, E. E., and H. T. Gier.** 1978. Lungworms in the coyotes on the Great Plains. J. Wildl. Dis. 14:314-316.—About one-third of *Canis latrans* from western Oklahoma had one or both species of lungworm (*Filaroides osleri* and *Capillaria aerophilia*). Parasites. Canidae.

811 **Morrison, E. E., and H. T. Gier.** 1979. Parasitic infection of *Filaroides osleri, Capillaria aerophila* and *Spirocera lupi* in coyotes from the southwestern United States. J. Wildl. Dis. 15:557-559.—*Canis latrans* from western Oklahoma and northern Texas had relatively high incidence of helminth infection. Parasites. Canidae.

Morrison, J. A., *see* 223, 381, 382, 383, 384.

812 **Morse, R. C.** 1957. The taxonomic status of two isolated populations of vespertilionid bats (*Antrozous* and *Pipistrellus*) in western Oklahoma. M.S. Thesis, Oklahoma State Univ., Stillwater. 43 pp.—New subspecies *Pipistrellus hesperus oklahomae* named with type locality in Greer County. *Antrozous bunkeri* relegated to subspecific status as *A. pallidus bunkeri.* Systematics, Distribution, Morphology. Vespertilionidae.

813 **Morse, R. C., and B. P. Glass.** 1960. The taxonomic status of *Antrozous bunkeri.* J. Mammal. 41:10-15.—*Antrozous bunkeri* (=*pallidus*) in Cimarron County, Oklahoma, and Barber County, Kansas. Systematics, Distribution. Vespertilionidae.

Morse, R. C., *see* 424.

814 **Moser, B.** 1974. Why hunt doe deer? Outdoor Okla. 30(9):11-13.—*Odocoileus virginianus.* Account for the layman. Ecology, Populations, Wildlife management, Popular. Cervidae.

Moulthrop, P. N., *see* 169.

815 **Moulton, M. P., J. R. Choate, and S. J. Bissell.** 1983. Biogeographic relationships of pocket gophers in southeastern Colorado. Southwest. Nat. 28:53-60.—Examined associations of plains pocket gopher (*Geomys bursarius*), yellow-faced pocket gopher (*Pappogeomys* [=*Cratogeomys*] *castanops*), and Botta's pocket gopher (*Thomomys bottae*). Study area included portions of Cimarron County. History, Distribution, Natural history, Ecology. Geomyidae.

816 **Murie, O. J.** 1951. The elk of North America. Stackpole Co., Harrisburg, Pennsylvania, and Wildlife Management Institute, Washington, D.C. 376 pp.—*Cervus elaphus* in Oklahoma occurred primarily in Wichita Mountains. Systematics, History, Distribution, Natural history, Behavior, Ecology, Populations, Parasites, Morphology, Wildlife management, Popular. Cervidae.

817 **Murphy, J.** 1984. Bats. Outdoor Okla. 40(4):40-45.—Popular article on Oklahoma's bats. Systematics, Distribution, Natural history, Behavior, Ecology, Morphology, Economics, Popular. Vespertilionidae, Molossidae.

818 **Murphy, J.** 1984. The crafty coyote. Outdoor Okla. 40(5):14-21.—Popular article on *Canis latrans* in Oklahoma. History, Distribution, Natural history, Behavior, Ecology, Populations, Economics, Wildlife management. Canidae.

819 **Murphy, J.** 1986. Oklahoma's "other" deer. Outdoor Okla. 42(6):14-17.—Popular article on Oklahoma mule deer (*Odocoileus hemionus*). Distribution, Natural history, Behavior, Ecology, Wildlife management, Popular. Cervidae.

820 **Murphy, M. F.** 1950. The ecology of the Osage woodrat, *Neotoma floridana osagensis* Blair, with emphasis on the helminths of the host. M.S. Thesis, Oklahoma State Univ., Stillwater. 50 pp.—Study assessed ectoparasites and helminths of *Neotoma floridana.* Described new species and host records for helminths. Distribution, Natural history, Behavior, Ecology, Parasites, Techniques, Economics. Cricetidae.

821 **Murphy, M. F.** 1952. Ecology and helminths of the Osage wood rat, *Neotoma floridana osagensis,* including the description of *Longistriata neotoma* n.sp. (Trichostrongylidae). Am. Midl. Nat. 48:204-218.—*Neotoma floridana osagensis* and *Mus musculus* in Payne County. Systematics, Distribution, Natural history, Behavior, Ecology, Parasites, Morphology. Cricetidae, Muridae.

822 **Murray, G. B.** 1977. Effects of environmental structure on the burrow distribution of thirteen-lined ground squirrels, *Spermophilus tridecemlineatus* (Sciuridae). M.S. Thesis, Univ. Oklahoma, Norman. 16 pp.—*Spermophilus tridecemlineatus* in Cleveland and Oklahoma counties. Burrow entrances were in areas with greater amounts of horizon obstructed and were near the obstructions. Ecology. Sciuridae.

823 **Murray, G. B., and B. M. Vestal.** 1979. Effects of environmental structure on the burrow distribution of thirteen-lined ground squirrels, *Spermophilus tridecemlineatus* (Sciuridae). Southwest. Nat. 24:79-86. —*Spermophilus tridecemlineatus* burrows in Oklahoma County and Cleveland County placed preferentially close to visual obstructions. Behavior. Sciuridae.

Mursaloglu, B., *see* 648.

824 **Myers, A. J.** 1965. Late Wisconsinan date for the Bar M local fauna. Okla. Geol. Notes 25:168-170.—Brief mention of *Dasypus bellus* from the Harper County site. Paleontology. Dasypodidae.

Myers, A. J., *see* 675.

825 **Myers, R. F.** 1964. Ecology of three species of myotine bats in the Ozark plateau. Ph.D. Dissertation, Univ. Missouri, Columbia. 250 pp.—Study of *Myotis lucifugus, M. sodalis,* and *M. grisescens.* Part of work done in caves in northeastern Oklahoma. Distribution, Natural history, Behavior, Ecology, Populations, Wildlife management. Vespertilionidae.

826 **Neel, C.** 1982. A bison skull from Logan County. Okla. Anthropol. Soc. Newsl. 30(5):6-9.—*Bison* recovered from bank of Gar Creek. Paleontology. Bovidae.

827 **Neel, C. D.** 1984. Ayers, 34Ma-126: A woodland period seasonal camp in the Cross Timbers, south-central Oklahoma. Contributions to Cross Timbers prehistory (P. L. Kawecki and D. G. Wyckoff, eds.). Okla. Archeol. Surv., Studies in Oklahoma's Past. No. 12:103-147.—Faunal remains included 308 fragments including those of several mammals. Paleontology. Castoridae, Canidae, Cervidae.

828 **Neel, C. D.** 1985. The Allison/Menifee mammoth excavation: Evidence for a Late Pleistocene stream course in central Oklahoma. Okla. Geol. Notes 45:188-204.—Site in northwestern Oklahoma City where *Mammuthus* sp. tusks found is described and analyzed in considerable detail. Paleontology, Ecology. Elephantidae.

Nejtek, M. E., *see* 1206.

Nelson, A. L., *see* 747.

829 **Nelson, E. W.** 1909. The rabbits of North America. N. Am. Fauna 29:1-314.—Systematics, History, Distribution, Natural history, Behavior, Ecology, Populations, Parasites, Genetics, Morphology, Economics. Leporidae.

830 **Nelson, E. W.** 1925. Status of the pronghorned antelope, 1922-1924. U.S. Dep. Agric. Bull. 1346:1-64.—Distribution of *Antilocapra americana* included Cimarron and Comanche counties. History, Distribution, Natural history, Populations, Morphology, Techniques, Economics, Wildlife management. Antilocapridae.

831 **Nelson, E. W., and E. A. Goldman.** 1930. Six new raccoons of the *Procyon lotor* group. J. Mammal. 11:453-459.—Distribution of *Procyon lotor hirtus,* new subspecies, includes Oklahoma. Systematics. Procyonidae.

832 **Nelson, J. S.** 1984. White-tailed deer and cattle interactions in southeastern Oklahoma. M.S. Thesis, Oklahoma State Univ., Stillwater. 65 pp.—Data on *Odocoileus virginianus* habitat use, activity patterns, home ranges, and food preferences. Natural history, Behavior, Ecology, Economics, Wildlife management. Cervidae.

833 **Nelson, K.** 1984. Genetic interactions between hybridizing cytotypes of the white-footed mouse, *Peromyscus leucopus.* M.S. Thesis, Texas Tech Univ., Lubbock. 65 pp.—Evaluation of intergrade zone in central Oklahoma. Systematics, Distribution, Genetics. Cricetidae.

834 **Neuman, R. W.** 1983. The buffalo in southeastern United States post-Pleistocene prehistory. Southeastern natives and their pasts. Okla. Archeol. Surv., Studies in Oklahoma's Past. No. 11:261-280.—Discussion of available archeological data on presence and utilization of bison (*Bison*) in prehistory of southeastern United States, including Oklahoma. Paleontology. Bovidae.

Newman, C., *see* 361.

Nice, L. B., *see* 1071.

835 **Nickel, P. A., and M. F. Hansen.** 1967. Helminths of bats collected in Kansas, Nebraska, and Oklahoma. Am. Midl. Nat. 78:481-486.—Trematodes, cestodes, and nematodes taken from a collection including *Myotis lucifugus, M. keenii, M. velifer, M. grisescens, Pipistrellus subflavus, Eptesicus fuscus,* and *Tadarida brasiliensis. Myotis grisescens* was

most frequently and most heavily parasitized. Parasites. Vespertilionidae, Molossidae.

Nieves, M. A., *see* 607, 608.

836 **Nowak, R.** 1974. Red wolf our most endangered mammal. Natl. Parks Mag. 48(8):9-12.—*Canis rufus* historical distribution included portions of Oklahoma. Systematics, History, Distribution, Natural history, Populations. Canidae.

837 **Nowak, R. M.** 1979. North American Quaternary *Canis.* Univ. Kans. Mus. Nat. Hist. Monogr. 6:1-154.—Includes specimens of *Canis rufus, C. latrans,* and *C. dirus* from Oklahoma. Systematics, Distribution, Morphology. Canidae.

838 **Nowak, R. M., and J. L. Paradiso.** 1983. Walker's mammals of the world. Johns Hopkins Univ. Press, Baltimore. 2 vols. 1,362 pp.—Fourth edition of work initiated by J. S. Walker. Discussions of families and genera with checklists of species. Systematics, History, Distribution, Natural history, Behavior, Ecology, Morphology, Economics, Bibliographies. Mammalia.

Nowak, R. M., *see* 859, 860.

839 **O'Connor, J.** 1977. The big game animals of North America. Outdoor Life, New York. 238 pp.—Distribution, Natural history, Behavior, Ecology, Morphology, Popular. Ursidae, Cervidae, Antilocapridae, Bovidae.

840 **O'Gara, B. W.** 1978. *Antilocapra americana.* Mamm. Species 90:1-7.—Distribution of *Antilocapra americana* includes portions of Oklahoma Panhandle. Summary of information in literature. Systematics, Paleontology, History, Distribution, Natural history, Behavior, Ecology, Populations, Genetics, Morphology, Physiology, Bibliographies. Antilocapridae.

841 **O'Meilia, M. E.** 1980. Competition between prairie dogs and beef cattle for range forage. M.S. Thesis, Oklahoma State Univ., Stillwater. 33 pp.—In Harper County study, *Cynomys ludovicianus* significantly reduced forage available to cattle. Pastures with prairie dogs had more small mammals and fewer insects than those without. Natural history, Behavior, Ecology, Populations, Economics, Wildlife management. Leporidae, Sciuridae, Geomyidae, Heteromyidae, Cricetidae, Bovidae.

842 **O'Meilia, M. E., F. L. Knopf, and J. C. Lewis.** 1982. Some consequences of competition between prairie dogs and beef cattle. J.

Range Manage. 35:580-585.—Study of effects of *Cynomys ludovicianus* on forage and cattle production in Woodward County. Also includes comments on *Sylvilagus audubonii, Geomys bursarius,* and seven species of small rodents. Ecology, Populations, Economics. Leporidae, Sciuridae, Geomyidae, Heteromyidae, Cricetidae.

O'Shea, T. J., *see* 564.

843 **Ockenfels, R. A.** 1980. Habitat requirements of white-tailed deer in the post oak-blackjack oak habitat type. M.S. Thesis, Oklahoma State Univ., Stillwater. 73 pp.—*Odocoileus virginianus* habits and habitats assessed. Natural history, Behavior, Ecology, Techniques, Wildlife management. Cervidae.

844 **Ockenfels, R. A., and J. A. Bissonette.** 1982. Estimates of white-tailed deer activity levels in Oklahoma. Proc. Annu. Conf. Southeast. Assoc. Fish Wildl. Agencies 36:445-453.—Activity levels of *Odocoileus virginianus* estimated by track counts on three study sites in Cross Timbers region of Oklahoma. Natural history, Ecology, Behavior. Cervidae.

845 **Ockenfels, R. A., and J. A. Bissonette.** 1983. A track plot system to monitor habitat use. Proc. Annu. Conf. Southeast. Assoc. Fish Wildl. Agencies 37:173-181.—Evaluation of habitat use by white-tailed deer (*Odocoileus virginianus*). Technique used to evaluate biotelemetry data. Ecology, Techniques, Wildlife management. Cervidae.

846 **Ockenfels, R. A., and J. A. Bissonette.** 1984. Temperature-related responses in north-central Oklahoma white-tailed deer. Pages 64-67 *in* Deer in the Southwest: A workshop (P. R. Krausman and N. S. Smith, eds.). Arizona Coop. Wildl. Res. Unit, Univ. Ariz., Tucson. 131 pp.—*Odocoileus virginianus.* Behavior, Ecology, Physiology, Wildlife management. Cervidae.

Ojeda C., M. M., *see* 230.

Ormiston, B. G., *see* 695.

Orr, H. W., *see* 1161.

847 **Ortenburger, A. I.** 1926. Some bats from Oklahoma. Proc. Okla. Acad. Sci. 6:122.—*Pipistrellus subflavus subflavus, Nyctinomus* (=*Tadarida*) *mexicanus* (=*brasiliensis*), *Nycteris* (=*Lasiurus*) *borealis borealis,* and *Nycticius* (=*Nycticeius*) *humeralis.* Account of specimens from McCurtain, Cleveland, and Pushmataha counties. Distribution. Vespertilionidae, Molossidae.

848 **Osborn, B.** 1941. Prairie dogs in shinnery (oak scrub) savannah. Proc. Okla. Acad. Sci. 21:97.—*Cynomys ludovicianus* in Roger Mills County colonizing shinnery savannah from adjacent short- and mixed-grass area. Distribution, Ecology. Sciuridae.

849 **Osborn, B.** 1942. Prairie dogs in shinnery (oak scrub) savannah. Ecology 23:110-115.—*Cynomys ludovicianus* in Roger Mills County occurred in *Quercus* savannah area. Distribution, Ecology. Sciuridae.

850 **Osborn, B., and P. F. Allen.** 1949. Vegetation of an abandoned prairie-dog town in tall grass prairie. Ecology 30:322-332.—Discussion of causes and effects of declining *Cynomys ludovicianus* populations. Natural history, Ecology, Populations. Sciuridae.

851 **Osborn, B., and W. H. Kellogg.** 1943. Wildlife occurrence and habitat conditions in Roger Mills and Custer counties, Oklahoma. Proc. Okla. Acad. Sci. 23:41-46.—Study includes 15 mammal species. Correlations between frequency of species and habitat elements. Ecology, Populations. Mammalia.

Osborn, B., *see* 8.

852 **Osgood, W. H.** 1900. Revision of pocket mice of the genus *Perognathus.* N. Am. Fauna 18:1-72.—Subgenera *Perognathus* and *Chaetodipus* both occur in western Oklahoma. Systematics, History, Distribution, Natural history, Behavior, Ecology, Morphology. Heteromyidae.

853 **Osgood, W. H.** 1909. Revision of the mice of the American genus *Peromyscus.* N. Am. Fauna 28:1-285.—Study of all recognized forms of *Peromyscus.* Systematics, History, Distribution, Natural history, Behavior, Ecology, Morphology, Economics. Cricetidae.

Owen, R. D., *see* 961.

854 **Owings, D. H., and W. J. Loughry.** 1985. Variation in snake-elicited jump-yipping by black-tailed prairie dogs: Ontogeny and snake-specificity. Z. Tierpsychol. 70:177-200.—Study of *Cynomys ludovicianus* conducted at Wichita Mountains National Wildlife Refuge near Cache, Comanche County. Natural history, Behavior. Sciuridae.

855 **Owings, D. H., and S. C. Owings.** 1979. Snake-directed behavior by black-tailed prairie dogs (*Cynomys ludovicianus*). Z. Tierpsychol. 49:35-54.—Study conducted in Wichita Mountains Wildlife Refuge near Cache, Comanche County and at Oklahoma City Zoo. Natural history, Behavior. Sciuridae.

Owings, S. C., *see* 855.

856 **Packard, R. L.** 1969. Taxonomic review of the golden mouse, *Ochrotomys nuttalli.* Univ. Kans. Mus. Nat. Hist. Misc. Publ. 51:373-406.—Distribution of *Ochrotomys nuttalli flammeus* included eastern Oklahoma. Systematics, Distribution, Morphology. Cricetidae.

Packard, R. L., *see* 474, 724.

857 **Palmer, D. B., Jr., and W. J. Gunier.** 1975. A preliminary survey of arthropods associated with bats and bat caves in Missouri and two counties of Oklahoma. J. Kans. Entomol. Soc. 48:524-531.—*Myotis grisescens* in Delaware County, and *Antrozous pallidus* and *Tadarida brasiliensis* in Woodward County. Distribution, Parasites. Vespertilionidae, Molossidae.

858 **Palmer, R. S.** 1954. The mammal guide. Doubleday and Co., Garden City, New Jersey. 384 pp.—Includes range maps showing approximate distributions in Oklahoma. Distribution, Natural history, Behavior, Ecology, Morphology, Techniques, Economics. Mammalia.

859 **Paradiso, J. L., and R. M. Nowak.** 1971. A report on the taxonomic status and distribution of the red wolf. U.S. Fish Wildl. Serv. Spec. Sci. Rep. Wildl. 145:1-36.—Lists historical and current distributions of *Canis rufus* and *C. latrans.* Systematics, History, Distribution, Populations, Morphology, Wildlife management. Canidae.

860 **Paradiso, J. L., and R. M. Nowak.** 1972. *Canis rufus.* Mamm. Species 22:1-4.—*Canis rufus* formerly distributed in eastern Oklahoma. Summary of information in literature. Systematics, Paleontology, History, Distribution, Natural history, Behavior, Ecology, Populations, Parasites, Genetics, Morphology, Economics, Bibliographies. Canidae.

Paradiso, J. L., *see* 838.

861 **Parker, R. L.** 1952. Ecology and economics of the western fox squirrel, *Sciurus niger rufiventer* (Geoffroy), in Payne County, Oklahoma. M.S. Thesis, Oklahoma State Univ., Stillwater. 84 pp.—Assessment of habits and habitats of *Sciurus niger* in Payne County. Natural history, Behavior, Ecology, Populations, Parasites, Economics, Wildlife management. Sciuridae.

862 **Parmalee, P. W.** 1980. Notes on the animal food resources of a Choctaw family in eastern Oklahoma. Pages 333-341 *in* The Eufala Lake project, a cultural resource survey and assessment (G. Perino, and J. Caffey, eds.). Museum of the Red River, Idabel, Oklahoma. 341 pp. —Almost 90% of mammal remains were of domestic species (cows and pigs). Paucity of remains of native species suggests that little ef-

fort was expended to obtain these animals. Paleontology. Leporidae, Sciuridae, Cricetidae, Procyonidae, Mustelidae, Cervidae.

863 **Parmalee, P. W., and A. E. Bogan.** 1981. Animal remains from the Roden site (Nlc 215), McCurtain County, Oklahoma. Pages 87-94 *in* Archeological investigations at the Roden site in McCurtain County, Oklahoma (G. Perino, ed.). Museum of the Red River, Publ. No. 1. Potsherd Press, Idabel, Oklahoma.—Total of 3,400 vertebrate remains recovered at site. At least 18 mammal species included. Paleontology. Didelphidae, Talpidae, Leporidae, Sciuridae, Geomyidae, Castoridae, Cricetidae, Canidae, Procyonidae, Mustelidae, Cervidae, Bovidae.

Parmalee, P. W., *see* 677.

Parton, W. J., *see* 913.

864 **Pascal, D. D., Jr., G. S. Jones, and J. O. Whitaker, Jr.** 1976. Mammals in the Indiana State University vertebrate collections. Indiana State Univ. Vertebr. Coll. Publ. 1:1-8.—Oklahoma specimens include *Blarina carolinensis* (=*hylophaga*), *Reithrodontomys montanus, Peromyscus maniculatus, P. leucopus,* and *Sigmodon hispidus.* Distribution. Soricidae, Cricetidae.

865 **Patrick, C. D., and J. A. Hair.** 1975. Ecological observations on *Dermacentor albipictus* (Packard) in eastern Oklahoma (Acarina: Ixodidae). J. Med. Entomol. 12:393-394.—*Cervus canadensis* (=*elaphus*) in Cherokee County. Ecology of winter ticks. Parasites. Cervidae.

866 **Patrick, C. D., and J. A. Hair.** 1978. White-tailed deer utilization of three different habitats and its influence on lone star tick populations. J. Parasitol. 64:1100-1106.—*Odocoileus virginianus* in eastern Oklahoma may modify habitat selection seasonally according to tick abundance. Behavior, Parasites. Cervidae.

867 **Patterson, B. D.** 1984. Geographic variation and taxonomy of Colorado and Hopi chipmunks (genus *Eutamias*). J. Mammal. 65:442-456.—Study includes one specimen of *Eutamias* (=*Tamias*) *quadrivittatus* from Cimarron County. Systematics, Distribution. Sciuridae.

Patterson, R. R., *see* 122.

Patton, C. S., *see* 222.

Patton, J. C., *see* 86.

Payne, L., *see* 1111.

868 **Peden, D. G., and G. J. Kraay.** 1979. Comparison of blood characteristics in plains bison, wood bison, and their hybrids. Can. J. Zool. 57:1778-1784.—Study includes *Bison bison* from Wichita Mountains Wildlife Refuge, Comanche County. Systematics, Genetics. Bovidae.

869 **Penman, J. T.** 1974. Bison discovery. Okla. Anthropol. Soc. Newsl. 22(5)7.—*Bison* from Love County. Paleontology. Bovidae.

870 **Penman, J. T.** 1974. Faunal analysis of the Hodge site, Custer County, Oklahoma. Pages 69-72 *in* Reports of the ARKLA salvage project (J. L. Hoffman, L. Neal, J. T. Penman, T. Prewitt, R. S. Saunders, C. S. Wallis, Jr., and D. G. Wycoff, eds.). Okla. Archeol. Surv. Contract Archeol. Ser. No. 1. 263 pp.—Six mammals found: eastern mole (*Scalopus aquaticus*), plains pocket gopher (*Geomys bursarius*), hispid cotton rat (*Sigmodon hispidus*), cottontail (*Sylvilagus* sp.), black-tailed jackrabbit (*Lepus californicus*), and bison (*Bison bison*). Paleontology, Talpidae, Leporidae, Geomyidae, Cricetidae, Bovidae.

871 **Penman, J. T.** 1975. Faunal interpretation. Pages 270-273 *in* Fort Washita from past to present: An archaeological report (K. E. Lewis, ed.). Okla. Histo. Soc. Ser. Anthropol. 1:1-287.—A black-footed ferret (*Mustela nigripes*) tentatively identified. Eastern cottontail (*Sylvilagus floridanus*), swamp rabbit (*S. aquaticus*), eastern woodrat (*Neotoma floridana*), gray fox (*Urocyon cinereoargenteus*), domestic pig (*Sus*), white-tailed deer (*Odocoileus virginianus*), and bison (*Bison bison*) also found. Paleontology. Leporidae, Cricetidae, Canidae, Mustelidae, Cervidae, Bovidae.

872 **Penman, J. T.** 1979. Faunal remains from the Pate site, Oklahoma. Bull. Okla. Anthropol. Soc. 28:147-150.—Six mammals recovered: eastern cottontail (*Sylvilagus floridanus*), hispid cotton rat (*Sigmodon hispidus*), eastern woodrat (*Neotoma floridana*), dog or coyote (*Canis* sp.), white-tailed deer (*Odocoileus virginianus*), and bison (*Bison bison*). Paleontology. Leporidae, Cricetidae, Canidae, Cervidae, Bovidae.

873 **Penney, D. F., and E. G. Zimmerman.** 1976. Genic divergence and local population differentiation by random drift in the pocket gopher genus *Geomys*. Evolution 30:473-483.—Includes samples of *Geomys bursarius* from two Oklahoma localities. Genetics. Geomyidae.

Penney, D. F., *see* 177.

Pennington, R. E., *see* 973.

874 **Perry, A. E.** 1965. Population analysis of the guano bat *Tadarida brasiliensis mexicana* (Saussure) using the lens-weight method of age determination. Ph.D. Dissertation, Oklahoma State Univ., Stillwater. 60 pp.—Population parameters assessed for *Tadarida brasiliensis* from Greer, Major, Woodward and Woods counties. Natural history, Populations, Morphology, Techniques. Molossidae.

875 **Perry, A. E., and G. Rogers.** 1964. Predation by great horned owl (*Bubo virginianus*) on young Mexican free-tailed bats (*Tadarida brasiliensis mexicana*) in Major County, Oklahoma. Southwest. Nat. 9:205.—Young *Tadarida brasiliensis* eaten by *Bubo virginianus* in Major County. Natural history. Molossidae.

Peterson, B., *see* 718.

876 **Peterson, D., M. Moore, S. Svec, and J. D. Rogers.** 1982. Appendix A: Artifact descriptions. Pages 119-160 *in* Spiro archaeology: 1980 research (J. D. Rogers, ed.). Okla. Archeol. Surv., Studies in Oklahoma's Past. No. 9. 234 pp.—Animal bones recovered from House Mound 5 included at least five species of mammals. Paleontology. Sciuridae, Cricetidae, Canidae, Cervidae.

Phillips, C. J., *see* 242.

877 **Phillips, P. G.** 1935. Rodent distribution in overgrazed and normal grasslands. M.S. Thesis, Univ. Oklahoma, Norman. 80 pp.—Analysis of rabbit and rodent habitats and populations in Cleveland and McClain counties. Distribution, Natural history, Ecology, Populations. Mammalia.

878 **Phillips, P. G.** 1936. The distribution of rodents in overgrazed and normal grasslands of central Oklahoma. Ecology 17:673-679.—*Lepus californicus, Sylvilagus floridanus, Geomys bursarius, Spermophilus tridecemlineatus, Peromyscus* spp., and *Sigmodon hispidus* studied in Cleveland and McClain counties. Deer mice and jackrabbits more abundant in overgrazed areas. Cottontails and cotton rats preferred undisturbed areas. Pocket gophers preferred moderately grazed lands. Ecology. Leporidae, Sciuridae, Geomyidae, Cricetidae.

879 **Pilcher, B. K., and G. E. Wampler.** 1981. Hunting season movements of white-tailed deer on Fort Sill Military Reservation, Oklahoma. Proc. Annu. Conf. Southeast. Assoc. Fish Wildl. Agencies 35:142-148. —Study of diurnal and seasonal movements of *Odocoileus virginianus* in Comanche County. Behavior, Populations, Wildlife management. Cervidae.

Pillaert, E. E., *see* 318.

880 **Pimlott, D. H., and P. W. Joslin.** 1968. The status and distribution of the red wolf. Trans. N. Am. Wildl. Nat. Res. Conf. 33:373-389.—Historical range of *Canis rufus rufus* in central and northeastern Oklahoma, as well as of *C. r. gregoryi* in southeastern Oklahoma. Red wolves probably now extirpated from state. Systematics, Distribution. Canidae.

881 **Pitts, B.** 1969. Dating the Bartow mammoth. Okla. Anthropol. Soc. Newsl. 17(6):5.—Fossil found near Mooreland in 1964 dated at 11,900 B.P. (ca. 9,936 B.C.). Paleontology. Elephantidae.

882 **Pizzimenti, J. J.** 1975. Evolution of the prairie dog genus *Cynomys*. Occas. Pap. Mus. Nat. Hist. Univ. Kans. 39:1-73.—Includes Oklahoma specimens. Chromosomal and serum protein studies. Systematics, Paleontology, Distribution, Behavior, Parasites, Genetics, Morphology. Sciuridae.

883 **Pollard, L.** 1984. Return of the river otter. Outdoor Okla. 40(4):2-7.—Reintroduction of *Lutra canadensis*. Distribution, Natural history, Wildlife management, Popular. Mustelidae.

884 **Pollard, L.** 1986. Oklahoma's endangered species. Outdoor Okla. 42(5):14-19.—Briefly discusses status in Oklahoma of 13 species considered as federally endangered, including the following four mammals: gray myotis (*Myotis grisescens*); Indiana myotis (*M. sodalis*); Ozark big-eared bat (*Plecotus townsendii ingens*); black-footed ferret (*Mustela nigripes*). Comments on status, Oklahoma range, habitat, reasons for decline, and recovery efforts. History, Distribution, Natural history, Populations, Morphology, Popular. Vespertilionidae, Mustelidae.

885 **Polley, H. W., and S. L. Collins.** 1984. Relationships of vegetation and environment in buffalo wallows. Am. Midl. Nat. 112:178-186.—Analysis of vegetation in active and inactive bison (*Bison bison*) wallows. Ecology. Bovidae.

886 **Polley, H. W., and L. L. Wallace.** 1986. The relationship of plant species heterogeneity to soil variation in buffalo wallows. Southwest. Nat. 31:493-501.—Analysis of plant species found in bison (*Bison bison*) wallows in Comanche County. Ecology. Bovidae.

887 **Poole, A. J., and V. S. Schantz.** 1942. Catalog of the type specimens of mammals in the United States National Museum, including the Biological Surveys Collection. U.S. Natl. Mus. Bull. 178:1-705.—Type locality of *Geomys breviceps dutcheri* in Muskogee County, *Dipodomys oklahomae* (=*ordii*) in Cleveland County, *Onychomys leucogaster*

breviauritus in Canadian County, *Microtus pinetorum nemoralis* in Adair County, and *Canis frustror* (=*latrans*) on Cimarron River about 100 miles west of Ft. Gibson, Oklahoma. Systematics, Distribution. Geomyidae, Heteromyidae, Cricetidae, Canidae, Mammalia.

Ports, M., *see* 211, 212.

Powell, J., *see* 384.

888 **Pratt, H. D., and N. E. Good.** 1954. Distribution of some common domestic rat ectoparasites in the United States. J. Parasitol. 40:113-129.—Several species of fleas, lice, and mites recorded from *Rattus norvegicus* or *R. rattus* in Oklahoma. Parasites. Muridae.

889 **Presnall, C. C.** 1958. The present status of exotic mammals in the United States. J. Wildl. Manage. 22:45-50.—Small herd of *Dama dama* in Muskogee County. Distribution. Cervidae.

890 **Preston, J. R., and R. E. Martin.** 1963. A gray shrew population in Harmon County, Oklahoma. J. Mammal. 44:268-270.—*Notiosorex c. crawfordi* in Harmon County. Distribution. Soricidae, Heteromyidae, Cricetidae.

Preston, J. R., *see* 751.

891 **Price, P. K., and M. L. Kennedy.** 1980. Genic relationships in the white-footed mouse, *Peromyscus leucopus,* and the cotton mouse, *Peromyscus gossypinus.* Am. Midl. Nat. 103:73-82.—Study includes *Peromyscus leucopus* from Dewey and Canadian counties. Systematics, Genetics. Cricetidae.

Price, R. D., *see* 1088.

892 **Puckette, W. L.** 1976. Notes on the occurrence of the short-faced bear (*Arctodus*) in Oklahoma. Proc. Okla. Acad. Sci. 56:67-68.—*Arctodus simus.* Paleontology, Distribution. Ursidae.

893 **Pursglove, S. R., Jr.** 1976. *Eucyathostomum webbi* sp. n. (Strongyloidea: Cloacinidae) from white-tailed deer (*Odocoileus virginianus*). J. Parasitol. 62:574-578.—*Odocoileus virginianus* from Pushmataha County, Oklahoma and McIntosh County, Georgia. Problems in taxonomy of genus *Eucyathostomum* discussed. Description of *Eucyathostomum webbi,* new species. Parasites. Cervidae.

894 **Pursglove, S. R., Jr.** 1977. Helminth parasites of white-tailed deer *Odocoileus virginianus* from New Jersey and Oklahoma. Proc. Helminthol. Soc. Wash. 44:107-108.—*Odocoileus virginianus* from Adair

and Pushmataha counties. Fifteen species of helminths found. Parasites. Cervidae.

895 **Pyle, R. A., and W. Caire.** 1979. Second record of *Tadarida macrotis* (Gray) (Chiroptera: Molossidae) in Oklahoma. Southwest. Nat. 24:389.—*Tadarida macrotis* in Oklahoma County. Distribution. Molossidae.

896 **Radovsky, F. J., and D. P. Furman.** 1963. The North American species of *Steatonyssus* (Acarina: Dermanyssidae). Ann. Entomol. Soc. Am. 56:268-276.—*Steatonyssus* mites occurred on *Antrozous pallidus, Eptesicus fuscus,* and *Nycticeius humeralis* in Oklahoma. Distribution, Parasites. Vespertilionidae.

897 **Rainey, D. G.** 1956. Eastern woodrat, *Neotoma floridana:* Life history and ecology. Univ. Kans. Publ. Mus. Nat. Hist. 8:535-646.—Includes map with Oklahoma distribution of *Neotoma floridana.* Distribution, Natural history, Behavior, Ecology, Populations, Parasites, Morphology, Physiology. Cricetidae.

898 **Raun, G. G.** 1962. A bibliography of the Recent mammals of Texas. Bull. Tex. Mem. Mus. 3:1-81.—Useful early bibliography (organized taxonomically and with some general subjects) of literature prior to 1960 relating to mammals in Texas. Bibliographies. Mammalia.

899 **Ray, C. E., and A. E. Sanders.** 1984. Pleistocene tapirs in the eastern United States. Pages 283-315 *in* Contributions in Quaternary vertebrate paleontology: A volume in memorial to John E. Guilday (H. H. Genoways and M. R. Dawson, eds.). Carnegie Mus. Nat. Hist. Spec. Publ. No. 8. 538 pp.—Cites Oklahoma locality (Holloman gravel pit, Tillman County) for *Tapirus haysii.* Systematics, Paleontology, Morphology. Tapiridae.

Ray, C. E., *see* 397.

Ray, D. E., *see* 577.

900 **Ray, M. A.** 1965. A report on excavations at the Pohly site, My-54, in northeastern Oklahoma. Bull. Okla. Anthropol. Soc. 18:1-68.—Total of 6,814 animal bones found of which 238 were bison (*Bison bison*), 5,282 were white-tailed deer (*Odocoileus virginianus*), and the rest were mostly small animals. Pronghorn (*Antilocapra americana*) also identified. Paleontology. Cervidae, Antilocapridae, Bovidae, Mammalia.

Read, J. A., *see* 956.

Reebs, J. E., *see* 788.

901 **Reese, J. B.** 1976. Effects of dehydration on metabolic rate and fluid distribution in the jackrabbit, *Lepus californicus.* M.S. Thesis, Univ. Oklahoma, Norman. 33 pp.—*Lepus californicus.* Includes specimens from Oklahoma (near Kingfisher) and New Mexico. Distribution, Physiology. Leporidae.

Reisen, N. T., *see* 903.

902 **Reisen, W. K., and T. L. Best.** 1973. An ectoparasite survey of *Dipodomys ordii* (Rodentia: Heteromyidae) from central Oklahoma. Ann. Entomol. Soc. Am. 66:599-601.—Eight ectoparasite species found in a year-long study of *Dipodomys ordii oklahomae* near Norman, Cleveland County. Parasites. Heteromyidae.

903 **Reisen, W. K., M. L. Kennedy, and N. T. Reisen.** 1976. Winter ecology of ectoparasites collected from hibernating *Myotis velifer* (Allen) in southwestern Oklahoma (Chiroptera: Vespertilionidae). J. Parasitol. 62:628-635.—*Myotis velifer, Plecotus townsendii* and *Pipistrellus subflavus* in Harmon County. Parasite genera collected were *Trichobius, Mydopsylla, Macronyssus, Paraspinturnix, Spinturnix, Ornithodoros, Albeckia, Nycteriglyphus,* and *Olabidocarpus.* Discusses parasite associations. Parasites. Vespertilionidae.

904 **Renaud, E. B.** 1928. L'antiquite de l'homme dans l'Amerique du Nord. Anthropologie (Paris) 38:23-49.—Discussion of Tillman County site, which yielded *Mylodon, Trilophodon, Equus, Elephas, Glyptodon, Megatherium,* and other fossil mammals. Paleontology. Mammalia.

905 **Reynolds, R. E.** 1976. Evaluating beaver guards on restricted flow risers of Soil Conservation Service flood control impoundments in Oklahoma. M.S. Thesis, Oklahoma State Univ., Stillwater. 46 pp.—*Castor canadensis* distributed statewide, most abundantly in southeast. Study assesses various structures for prevention of plugging of impoundment outlets. Distribution, Behavior, Ecology, Populations, Wildlife management. Castoridae.

906 **Reynolds, R. E.** 1977. Distribution of beaver in Oklahoma. Proc. Okla. Acad. Sci. 57:83-85.—*Castor canadensis.* Distribution, Populations. Castoridae.

907 **Reynolds, R. E., and J. C. Lewis.** 1978. Evaluating beaver guards on restricted flow risers on flood control impoundments. Proc. Annu. Conf. Southeast. Assoc. Fish Wildl. Agencies 30:455-462.—Several methods tested to prevent *Castor canadensis* from plugging restricted-

flow risers in impoundments in eastern and southeastern Oklahoma. Wildlife management. Castoridae.

Rice, D. W., *see* 645.

908 **Richardson, D. M.** 1970. Adjustment of caloric intake by the kangaroo rat (*Dipodomys ordii richardsoni*) because of environmental temperature change. M.S. Thesis, Oklahoma State Univ., Stillwater. 42 pp.—*Dipodomys ordii* maintained consistent caloric intake at varying temperatures, but increased protein-to-oil ratio at lower temperatures. Behavior, Ecology, Physiology. Heteromyidae.

909 **Richardson, R. G.** 1958. The comparative behavior of some Oklahoma mice of the families Heteromyidae, Cricetidae, and Muridae, and its ecological implications. M.S. Thesis, Univ. Oklahoma, Norman. 49 pp.—Assessment of intraspecific and interspecific behavior of *Dipodomys ordii, Perognathus* (=*Chaetodipus*) *hispidus, Sigmodon hispidus, Peromyscus leucopus,* and *Mus musculus* from Cleveland, McClain, and Bryan counties. Behavior, Ecology. Heteromyidae, Cricetidae, Muridae.

910 **Riddle, B. R., and J. R. Choate.** 1986. Systematics and biogeography of *Onychomys leucogaster* in western North America. J. Mammal. 67:233-255.—Morphometric and karyotypic analysis of variation in northern grasshopper mouse. Includes specimens from Harper County, Oklahoma. Systematics, Paleontology, Distribution, Genetics, Morphology. Cricetidae.

911 **Ring, D. A.** 1968. Low population density and habitat preference of *Sigmodon.* M.S. Thesis, Oklahoma State Univ., Stillwater. 28 pp.—Study in Payne County examined fluctuations in abundance and habitat preferences of hispid cotton rat (*Sigmodon hispidus*). Natural history, Behavior, Ecology, Populations. Cricetidae.

912 **Rinker, G. C., and C. W. Hibbard.** 1952. A new beaver, and associated vertebrates, from the Pleistocene of Oklahoma. J. Mammal. 33:98-101.—*Paradipoides stovalli,* new genus and species, type locality in Beaver County. Associated mammals were *Castoroides, Mammuthus, Equus,* and a canid. Paleontology. Castoridae, Canidae, Equidae, Elephantidae.

913 **Risser, P. G., E. C. Birney, H. D. Blocker, W. May, W. J. Parton, and J. A. Wiens.** 1979. The true prairie ecosystem. Hutchinson Ross Publishing Co., Stroudsburg, Pennsylvania. 557 pp.—Includes data on Oklahoma mammals. Distribution, Ecology, Economics. Mammalia.

Risser, P. G., *see* 961.

Robbins, L. W., *see* 86, 100, 603.

Robinson, A., *see* 183.

914 **Roe, F. G.** 1951. The North American buffalo. A critical study of the species in its wild state. Univ. Toronto Press, Toronto. 957 pp.—General study of *Bison bison.* Paleontology, History, Distribution, Natural history, Populations, Popular. Bovidae.

Rogers, C. E., *see* 465.

Rogers, G., *see* 875.

915 **Rogers, G. C.** 1965. Aberrant coloration in *Myotis velifer incautus.* Southwest. Nat. 10:311.—A partially albino specimen of *Myotis velifer* from Vickery Cave, Major County. Natural history. Vespertilionidae.

916 **Rogers, G. C.** 1965. Studies of the incidence of gastric and intestinal helminths of *Tadarida brasiliensis mexicana* (Saussure). M.S. Thesis, Oklahoma State Univ., Stillwater. 53 pp.—Study of helminth parasites from populations of *Tadarida brasiliensis* in Major, Woods, and Woodward counties. Parasites, Techniques. Molossidae.

917 **Rogers, G. C.** 1972. Ecology of the Mexican freetailed bat, *Tadarida brasiliensis mexicana,* in Oklahoma and Texas. Ph.D. Dissertation, Oklahoma State Univ., Stillwater. 74 pp.—Study of reproductive and population aspects of *Tadarida brasiliensis* colonies in Greer, Major, Woods, and Woodward counties. Distribution, Natural history, Behavior, Ecology, Populations, Morphology. Molossidae.

918 **Rogers, J. D.** 1980. Spiro archeology: 1979 excavations. Okla. Archeol. Surv., Studies in Oklahoma's Past. No. 6. 223 pp.—Animal remains from site identified including five species of mammals. Paleontology. Didelphidae, Leporidae, Procyonidae, Felidae, Cervidae.

Rogers, J. D., *see* 876.

919 **Rohrbaugh, C.** 1971. Maysville mammoth investigations. Okla. Anthropol. Soc. Newsl. 19(7):2.—Mammoth uncovered while building municipal lake for Maysville. Paleontology. Elephantidae.

920 **Rohrbaugh, C. L.** 1972. The Stuckey mammoth. Okla. Anthropol. Soc. Newsl. 20(4) 5-6.—Mammoth (*Mammuthus*) found near Chattanooga, Oklahoma. Paleontology. Elephantidae.

921 **Rohwer, S. A., and D. L. Kilgore, Jr.** 1973. Interbreeding in the aridland foxes *Vulpes velox* and *V. macrotis.* Syst. Zool. 22:157-165.—Phenetic analysis involving 14 skull measurements of kit and swift

foxes (*Vulpes velox* and *V. macrotis*). Oklahoma specimens used. Systematics, Distribution. Canidae.

922 **Rolley, R. E.** 1983. Behavior and population dynamics of bobcats in Oklahoma. Ph.D. Dissertation, Oklahoma State Univ., Stillwater. 98 pp.—Study of *Felis rufus* population densities, age structure, mortality causes, home ranges, and prey preferences. Other mammals mentioned. Behavior, Populations, Techniques, Wildlife management. Leporidae, Felidae, Mammalia.

923 **Rolley, R. E.** 1985. Dynamics of a harvested bobcat population in Oklahoma. J. Wildl. Manage. 49:283-292.—*Felis rufus.* Demographic parameters estimated to determine population status. Populations, Economics, Wildlife management. Felidae.

924 **Rolley, R. E., and W. D. Warde.** 1985. Bobcat habitat use in southeastern Oklahoma. J. Wildl. Manage. 49:913-920.—Seven male and four female *Felis rufus* studied using radio tracking. Habitat use and availability determined from Landsat digital data. Habitat use differed between sexes and in relation to other factors. Lagomorph and cricetine rodent populations higher in clear-cut than in other habitats. Natural history, Ecology, Techniques, Wildlife management. Leporidae, Cricetidae, Felidae.

Rolley, R. E., *see* 427.

925 **Romer, A. S.** 1933. Pleistocene vertebrates and their bearing on the problem of human antiquity in North America. Pages 49-83 *in* The American aborigines (D. Jenness, ed.). Univ. Toronto Press, Toronto. —Entire volume republished in 1973 by Cooper Square Publishers, New York. Discussion of Early Pleistocene site in Tillman County in which artifacts are associated with mammal remains. Lists 10 large mammals found. Paleontology. Mammalia.

926 **Roosevelt, T.** 1905. A wolf hunt in Oklahoma. Scribner's Mag. 38:513-532.—*Canis latrans* and *Cynomys ludovicianus* noted as common in Kiowa and Comanche counties. Popular. Leporidae, Sciuridae, Canidae.

927 **Roosevelt, T., T. S. Van Dyke, D. G. Elliot, and A. J. Stone.** 1902. The deer family. Grosset and Dunlap, New York. 334 pp.—Distribution of *Odocoileus hemionus, O. virginianus* and *Antilocapra americana* include portions of Oklahoma. Distribution, Natural history. Cervidae, Antilocapridae.

Root, C. R., *see* 609.

928 **Roseberry, S. E.** 1980. Effects of sport hunting on raccoon reproduction, survival and behavior. M.S. Thesis, Oklahoma State Univ., Stillwater. 29 pp.—Comanche County study of *Procyon lotor* indicates that reproduction inversely related to population density, and adequately compensates for hunting and predation pressures. Natural history, Behavior, Ecology, Populations, Parasites, Morphology, Wildlife management. Procyonidae.

929 **Ross, A.** 1961. Biological studies on bat ectoparasites of the genus *Trichobius* (Diptera: Streblidae) in North America, north of Mexico. Wasmann J. Biol. 19:229-245.—In Oklahoma, *Trichobius major* reported from *Tadarida brasiliensis* and *Myotis velifer incautus.* Also, *Trichobius sphaeronotus* from *Tadarida brasiliensis.* Parasites. Vespertilionidae, Molossidae.

930 **Roudabush, R. L.** 1936. Arthropod and helminth parasites of the American bison (*Bison bison*). J. Parasitol. 22:517-518.—Examination of 13 *Bison bison* from Comanche County produced six species of parasites. Parasites. Bovidae.

Rouk, C. S., *see* 425.

931 **Rouse, C. B.** 1941. Notes of wintering foraging habits of antelope in Oklahoma. J. Mammal. 22:57-60.—*Antilocapra americana* stocked on Wichita Mountains Refuge. Ecology, Wildlife management. Antilocapridae.

932 **Rowe, D. J.** 1980. Spotlighting deer for determination of population trends in Oklahoma. M.S. Thesis, Oklahoma State Univ., Stillwater. 27 pp.—Frequency of observations of *Odocoileus virginianus* greater in oak-hickory forest than in Cross Timbers. Numbers unaffected by weather and traffic. Distribution, Ecology, Populations, Techniques, Wildlife management. Cervidae.

Royce, M. A., *see* 213.

933 **Ruffer, D. G.** 1963. A checklist of the mammals of the Oliver Wildlife Preserve, Cleveland County, Oklahoma. Proc. Okla. Acad. Sci. 44:66-68.—Fifteen species and their numbers recorded. Distribution, Ecology. Mammalia.

934 **Ruffer, D. G.** 1964. Studies on the ethology of the northern grasshopper mouse (*Onychomys leucogaster*). Ph.D. Dissertation, Univ. Oklahoma, Norman. 104 pp.—Brief taxonomic and thorough ethological analysis of *Onychomys leucogaster.* Systematics, Paleontology, Distribution, Natural history, Behavior, Ecology. Cricetidae.

935 **Ruffer, D. G.** 1965. Burrows and burrowing behavior of *Onychomys leucogaster.* J. Mammal. 46:241-247.—*Onychomys leucogaster* specimens from Woods, Beckham, and Harmon counties. Grasshopper mice dug nest burrows, retreat burrows, and miscellaneous burrows. Mentions *Dipodomys, Sigmodon, Perognathus, Peromyscus, Marmota,* and *Miochonys* (extinct genus). Paleontology, Natural history, Behavior, Ecology. Sciuridae, Heteromyidae, Cricetidae.

936 **Ruffer, D. G.** 1965. Juvenile molt of *Onychomys leucogaster.* J. Mammal. 46:338-339.—Northern grasshopper mouse (*Onychomys leucogaster*). Natural history. Cricetidae.

937 **Ruffer, D. G.** 1965. Sexual behavior of the northern grasshopper mouse (*Onychomys leucogaster*). Anim. Behav. 13:447-452.—Description of feeding, courtship, parental, and play behavior of *Onychomys leucogaster.* Natural history, Behavior. Cricetidae.

938 **Ruffer, D. G.** 1968. Agonistic behavior of the northern grasshopper mouse (*Onychomys leucogaster breviauritus*). J. Mammal. 49:481-487. —Specimens from Oklahoma localities. Mentions *Dipodomys, Reithrodontomys, Peromyscus, Sigmodon, Mus, Microtus,* and *Synaptomys.* Behavior. Heteromyidae, Cricetidae, Muridae.

Russell, D. E., *see* 945.

939 **Russell, R. J.** 1968. Evolution and classification of the pocket gophers of the subfamily Geomyinae. Univ. Kans. Publ. Mus. Nat. Hist. 16:473-579.—Includes Oklahoma specimens and holotypes. Systematics, Paleontology, History, Distribution, Morphology. Geomyidae.

940 **Russell, R. J.** 1968. Revision of pocket gophers of the genus *Pappogeomys.* Univ. Kans. Publ. Mus. Nat. Hist. 16:581-776.—Distribution of *Pappogeomys* (=*Cratogeomys*) *castanops* includes Oklahoma Panhandle. Systematics, Distribution, Natural history, Ecology, Morphology. Geomyidae.

941 **Sanborn, C. C.** 1947. Catalog of type specimens of mammals in Chicago Natural History Museum. Fieldiana Zool. 32:209-285.—Lists seven type specimens from Oklahoma. Systematics, Distribution. Soricidae, Talpidae, Leporidae, Sciuridae, Heteromyidae, Cricetidae.

Sanders, A. E., *see* 899.

942 **Sanders, I. T.** 1951. How to know the American mammals. Little, Brown, and Co., Boston. 164 pp.—Field guide. Distribution, Natural history, Behavior, Ecology, Morphology, Popular. Mammalia.

943 **Savage, D. E.** 1941. Two new Middle Pliocene carnivores from Oklahoma with notes on the Optima fauna. Am. Midl. Nat. 25:692-710. —Names *Vulpes stenagnathus* and *Felis proterolyncis.* Found representatives of 15 families of mammals. Paleontology, Distribution. Didelphidae, Megalonychidae, Leporidae, Mylagaulidae, Canidae, Ursidae, Mustelidae, Felidae, Mastodontidae, Equidae, Rhinocerotidae, Tayassuidae, Camelidae, Cervidae, Antilocapridae.

944 **Savage, D. E.** 1946. A mandible of *Megalonyx* from the Pleistocene of Oklahoma. J. Mammal. 27:388-390.—*Megalonyx jeffersonii oklahomensis* in Grant County. Paleontology. Megalonychidae.

945 **Savage, D. E., and D. E. Russell.** 1983. Mammalian paleofaunas of the world. Addison-Wesley Publishing Co., London. 432 pp.—Includes mammal data from a number of Oklahoma Miocene, Pliocene, and Pleistocene localities. Systematics, Paleontology, Distribution. Mammalia.

Savage, D. E., *see* 576.

Schaefer, K., *see* 1192.

Schantz, V. S., *see* 887.

946 **Schemnitz, S. D., and E. Ables.** 1962. Notes on the food habits of the great horned owl in western Oklahoma. Condor 64:328-329.—Great horned owls (*Bubo virginianus*) in Cimarron County ate primarily *Sylvilagus floridanus* and rodents. Natural history, Ecology. Leporidae, Geomyidae, Heteromyidae, Cricetidae, Mustelidae.

Schemnitz, S. D., *see* 339.

947 **Schendel, R. R.** 1940. Life history notes on *Sigmodon hispidus texianus* with special emphasis on populations and nesting habits. M.S. Thesis, Oklahoma A and M Coll., Stillwater. 40 pp.—Payne County study of hispid cotton rat (*Sigmodon hispidus*) indicated high densities followed by population crash. Natural history, Behavior, Ecology, Populations, Economics. Sciuridae, Cricetidae.

948 **Schendel, R. R.** 1942. Mammals of Payne County, Oklahoma. Proc. Okla. Acad. Sci. 22:75-76.—List of 47 mammal species for Payne County. Range records for *Urocyon cinereoargenteus ocythous, Lynx* (=*Felis*) *rufus, Perognathus* (=*Chaetodipus*) *hispidus spilotus, Castor canadensis,* and *Reithrodontomys montanus griseus.* Distribution. Mammalia.

949 **Scherman, B.** 1984. Elk monitoring. Okla. Dep. Wildl. Cons. Job 01, Study 01, W-80-R-23. 2 pp.—Elk (*Cervus elaphus*) sightings and herd

locations recorded for northeastern Oklahoma. Approximately 175 animals tabulated. Distribution, Populations, Wildlife management. Cervidae.

950 **Schmidly, D. J.** 1972. Geographic variation in the white-ankled mouse, *Peromyscus pectoralis*. Southwest. Nat. 17:113-138.—Includes specimens housed at Oklahoma State University. Geographic, individual, and secondary sexual variation. Ways to distinguish *Peromyscus pectoralis* from *P. boylii*. Recognizes three subspecies (*P. pectoralis pectoralis, P. p. laceianus,* and *P. p. collinus*). *Peromyscus pectoralis eremicoides* is synonym for *P. p. pectoralis*. Systematics, Distribution, Natural history, Ecology, Populations, Morphology. Cricetidae.

951 **Schmidly, D. J.** 1973. Geographic variation and taxonomy of *Peromyscus boylii* from Mexico and the southern United States. J. Mammal. 54:111-130.—*Peromyscus attwateri* is separate from *P. boylii* group. Numerical approach. Mexico, New Mexico, Texas, Oklahoma, Kansas, Arkansas, and Missouri. Systematics, Morphology. Cricetidae.

952 **Schmidly, D. J.** 1974. *Peromyscus attwateri*. Mamm. Species 48:1-3.—Distribution of *Peromyscus attwateri* includes central and eastern Oklahoma. Summary of information in literature. Systematics, Paleontology, Distribution, Natural history, Behavior, Ecology, Parasites, Genetics, Morphology, Physiology, Bibliographies. Cricetidae.

953 **Schmidly, D. J.** 1974. *Peromyscus pectoralis*. Mamm. Species 49:1-3.—Distribution of *Peromyscus pectoralis* includes Love County, Oklahoma. Summary of information in literature. Systematics, Paleontology, Distribution, Natural history, Behavior, Ecology, Parasites, Genetics, Morphology, Bibliographies. Cricetidae.

954 **Schmidly, D. J.** 1983. Texas mammals east of the Balcones Fault zone. Texas A&M Univ. Press, College Station. 400 pp.—Records of mammals in region adjacent to southeastern Oklahoma. Includes keys. Distribution, Natural history. Mammalia.

955 **Schmidly, D. J.** 1984. The furbearers of Texas. Tex. Parks Wildl. Dep. Bull. 111:1-550.—Information on furbearers in area adjacent and south of Oklahoma. Distribution, Natural history, Behavior, Ecology. Mammalia.

956 **Schmidly, D. J., and J. A. Read.** 1986. Cranial variation in the bobcat (*Felis rufus*) from Texas and surrounding states. Occas. Pap. Mus.

Tex. Tech Univ. 101:1-39.—Oklahoma specimens included from 30 counties, along with those from Arkansas, Louisiana, Texas and New Mexico. Systematics, Morphology. Felidae.

Schmidly, D. J., *see* 116, 602, 649, 704, 748, 1193, 1194, 1195.

957 **Schmidt, J. L., and D. L. Gilbert, eds.** 1978. Big game of North America: Ecology and management. Stackpole Books, Harrisburg, Pennsylvania. 494 pp.—Includes broad coverage of big game species, including several found in Oklahoma. Little information specific to Oklahoma. Natural history, Wildlife management. Ursidae, Felidae, Cervidae.

Schmidt-Nielsen, K., *see* 463.

958 **Schmitt, K., and R. Toldan, Jr.** 1953. The Brown site, Gd-1, Grady County, Oklahoma. Bull. Tex. Archaeol. Paleontol. Soc. 24:141-176. —Excavation of Grady County site produced bones of bison, deer, and rabbits. Paleontology. Leporidae, Cervidae, Bovidae.

959 **Schneider, F. E.** 1969. The Roy Smith site, Bv-14, Beaver County, Oklahoma. Bull. Okla. Anthropol. Soc. 18:119-179.—Faunal remains included at least 12 bison (*Bison bison*), 3 white-tailed deer (*Odocoileus virginianus*), 1 pronghorn (*Antilocapra americana*), 1 black-tailed jackrabbit (*Lepus californicus*), 1 prairie dog (*Cynomys* sp.), 1 *Reithrodontomys* sp., and 1 *Peromyscus* sp. Paleontology. Leporidae, Sciuridae, Cricetidae, Cervidae, Antilocapridae, Bovidae.

960 **Schnell, G. D., J. A. Grzybowski, D. J. Hough, and T. M. McKenna.** 1985. Evaluation of spatial patterning in Oklahoma furbearer populations. Southwest. Nat. 30:225-238.—Furbearer data from scent-post censuses and pelt harvest data analyzed for geographic patterning. Species include Virginia opossum (*Didelphis virginianus*), coyote (*Canis latrans*), red fox (*Vulpes vulpes*), gray fox (*Urocyon cinereoargenteus*), raccoon (*Procyon lotor*), striped skunk (*Mephitis mephitis*), and bobcat (*Felis rufus*), as well as domestic dogs and cats. Distribution, Populations, Techniques. Didelphidae, Canidae, Procyonidae, Mustelidae, Felidae.

961 **Schnell, G. D., R. D. Owen, R. K. Chesser, and P. G. Risser.** 1980. Populations of small mammals in north-central Oklahoma. Southwest. Nat. 25:67-80.—Populations estimated for grassland and woodland habitats. Substantial variation found among seasons and among habitats. Natural history, Populations. Soricidae, Sciuridae, Heteromyidae, Cricetidae, Muridae.

Schnell, G. D., *see* 132, 656, 1133.

962 Schoff, S. L. 1954. Pliocene and Pleistocene fossils from Beaver County, Oklahoma. Proc. Okla. Acad. Sci. 35:94.—Teeth of mastodon (*Mastodon serridentinus*) and horse (*Calippus*) from Pliocene of Beaver County. Paleontology. Mastodontidae, Equidae.

963 Schultz, C. B., and L. D. Martin. 1972. Two lynx-like cats from the Pliocene and Pleistocene. Bull. Univ. Nebr. State Mus. 9:197-203.—Indicates that *Lynx stouti* older than and perhaps ancestral to "*Felis*" *proterolyncis* from the Oklahoma Optima fauna. Systematics, Paleontology. Felidae.

964 Schultz, C. B., L. D. Martin, and R. G. Corner. 1975. Middle and Late Cenozoic tapirs from Nebraska. Bull. Univ. Nebr. State Mus. 10:1-21.—Includes distribution map for Pleistocene tapirs that shows two Oklahoma localities. Paleontology, Distribution. Tapiridae.

965 Schultz, G. E., and E. P. Cheatum. 1970. *Bison occidentalis* and assorted invertebrates from the Late Wisconsin of Randell County, Texas. J. Paleontol. 44:16-30.—Mention of *Bison occidentalis* from Oklahoma. Paleontology, Distribution. Bovidae.

966 Schwagmeyer, P. L. 1984. Multiple mating and intersexual selection in thirteen-lined ground squirrels. Pages 275-294 *in* The biology of ground-dwelling squirrels: Annual cycles, behavioral ecology, and sociality (J. O. Murie and G. R. Michener, eds.). Univ. Nebraska Press, Lincoln. 459 pp.—Studies of *Spermophilus tridecemlineatus* in Oklahoma. Natural history, Behavior. Sciuridae.

967 Schwagmeyer, P. L. 1986. Effects of multiple mating on reproduction in female thirteen-lined ground squirrels. Anim. Behav. 34:297-298.—Multiple-mated versus single-mated female *Spermophilus tridecemlineatus* in vicinity of Norman, Oklahoma compared with respect to incidence of pregnancy, incidence of litter loss prior to above-ground emergence, and mean reproductive success. Natural history, Behavior. Sciuridae.

968 Schwagmeyer, P. L., and S. J. Wootner. 1985. Mating competition in an asocial ground squirrel, *Spermophilus tridecemlineatus.* Behav. Ecol. Sociobiol. 17:291-296.—Study of Oklahoma thirteen-lined ground squirrels. Natural history, Behavior. Sciuridae.

969 Schwagmeyer, P. L., and S. J. Wootner. 1986. Scramble competition polygyny in thirteen-lined ground squirrels: The relative contributions of overt conflict and competitive mate searching. Behav. Ecol. Sociobiol. 19:359-364.—Overt conflict component of intrasexual competition in *Spermophilus tridecemlineatus* studied in Norman, Oklahoma. Natural history, Behavior. Sciuridae.

970 **Schwartz, C. W., and E. R. Schwartz.** 1959. The wild mammals of Missouri. Univ. Missouri Press and Missouri Conservation Commission, Columbia. 341 pp.—Records of mammals along portion of Oklahoma's eastern border. Distribution, Natural history. Mammalia.

971 **Schwartz, C. W., and E. R. Schwartz.** 1981. The wild mammals of Missouri. Revised edition. Univ. Missouri Press and Department of Conservation, Columbia, Missouri. 356 pp.—Includes range maps showing approximate distributions in Oklahoma. Distribution, Natural history, Behavior, Ecology, Morphology, Techniques, Economics. Mammalia.

Schwartz, E. R., *see* 970, 971.

Scott, C. M., *see* 1112.

972 **Sealander, J. A.** 1979. A guide to Arkansas mammals. River Road Press, Conway, Arkansas. 313 pp.—Systematics, Paleontology, History, Distribution, Natural history, Behavior, Ecology, Parasites, Morphology, Techniques, Economics, Wildlife management, Bibliographies. Mammalia.

973 **Segelquist, C. A., and R. E. Pennington.** 1968. Deer browse in the Ouachita Forest in Oklahoma. J. Wildl. Manage. 32:623-626.—Scarcity of browse may limit *Odocoileus virginianus* populations in southeastern Oklahoma. Natural history, Ecology, Techniques, Wildlife management. Cervidae.

Selander, R. K., *see* 88.

974 **Self, J. T., and T. J. McKnight.** 1950. Platyhelminths from the fur bearers in the Wichita Mountains Wildlife Refuge, with a special reference to *Oochoristica* spp. Am. Midl. Nat. 43:58-61.—*Spilogale interrupta* (=*gracilis*), *Mephitis varians* (=*mephitis*), *Didelphis virginiana, Canis latrans, Procyon lotor,* and *Felis catus* in Comanche County. *Taenia taeniaeformis, Diphylidium caninum, Mesocestoides latus, T. pisiformis, Rhophalias macrocanthus, Mesocestoides variabilis, Oochoristica mephitis,* and *C. oklahomensis.* Parasites. Didelphidae, Canidae, Mustelidae, Procyonidae, Felidae.

Self, J. T., *see* 1054, 1055.

975 **Sellards, E. H.** 1932. Geologic relations of deposits reported to contain artifacts at Frederick, Oklahoma. Geol. Soc. Am. Bull. 43:783-

796.—Lists genera reported from Tillman County deposits. Paleontology. Mammalia.

976 **Sellards, E. H.** 1940. Early man in America: Index to localities, and selected bibliography. Geol. Soc. Am. Bull. 51:373-432.—Includes discussion of sites containing mammalian fossils in Ottawa and Tillman counties. Paleontology, Bibliographies. Mammalia.

977 **Semtner, P. J., and J. A. Hair.** 1973. Distribution, seasonal abundance, and hosts of the Gulf coast tick in Oklahoma. Ann. Entomol. Soc. Am. 66:1264-1268.—Gulf Coast tick found in eastern Oklahoma on cattle and several wild mammal species. Parasites. Mammalia.

978 **Seton, E. T.** 1925. Lives of game animals. Vol. 1. Cats, wolves, and foxes. Doubleday, Doran and Co. Garden City, New York. 640 pp.—History, Distribution, Natural history, Behavior, Ecology, Morphology, Wildlife management, Popular. Canidae, Felidae.

979 **Seton, E. T.** 1926. Lives of game animals. Vol. 2. Bears, coons, badger, skunks, and weasels. Doubleday, Doran and Co. Garden City, New York. 740 pp.—History, Distribution, Natural history, Behavior, Ecology, Morphology, Wildlife management, Popular. Ursidae, Procyonidae, Mustelidae.

980 **Seton, E. T.** 1927. Lives of game animals. Vol. 3. Hoofed animals. Doubleday, Doran and Co. Garden City, New York. 780 pp.—History, Distribution, Natural history, Behavior, Ecology, Morphology, Wildlife management, Popular. Cervidae, Antilocapridae, Bovidae.

981 **Seton, E. T.** 1928. Lives of game animals. Vol. 4. Squirrels, rabbits, armadillo, and opossum. Doubleday, Doran and Co. Garden City, New York. 949 pp.—History, Distribution, Natural history, Behavior, Ecology, Morphology, Wildlife management, Popular. Didelphidae, Dasypodidae, Leporidae, Sciuridae.

982 **Setzer, H. W.** 1949. Subspeciation in the kangaroo rat, *Dipodomys ordii.* Univ. Kans. Publ. Mus. Nat. Hist. 1:473-573.—*Dipodomys ordii richardsoni* and *D. o. oklahomae* in Oklahoma. Systematics, Paleontology, Distribution, Natural history, Ecology, Populations, Genetics, Morphology. Heteromyidae.

983 **Shaeffer, J. B.** 1961. Recent mammoth activity of the salvage project. Okla. Anthropol. Soc. Newsl. 9(6):1-3.—Mammoth skeletons found in Kiowa, Blaine, and Osage counties. Paleontology. Elephantidae.

Shalaway, S., *see* 1095.

984 **Sharrock, F. W.** 1961. The Grant site of the Washita River focus. Bull. Okla. Anthropol. Soc. 9:1-66.—Appendix by D. Ahshapanek lists mammals identified. Paleontology, Distribution. Mammalia.

985 **Shaw, J. H.** 1986. Belligerent. Anim. Kingdom 89(2):44-47.—Bison (*Bison bison*) in Oklahoma. Natural history, Wildlife management, Popular. Bovidae.

Shaw, J. H., *see* 371, 544, 729.

986 **Shaw, M.** 1983. Big game disease research. Okla. Dep. Wildl. Cons. Job 01, Study 03, W-80-R-22. 5 pp.—Limited summary data on 20 fawns and 24 adult white-tailed deer (*Odocoileus virginianus*) taken in 1982. Natural history, Parasites, Wildlife management. Cervidae.

987 **Shaw, M.** 1985. Big game disease research. Okla. Dep. Wildl. Cons. Job 01, Study 03, W-80-R-24. 6 pp.—Limited summary data from 1 buck, 11 fawns, and 35 adult doe white-tailed deer (*Odocoileus virginianus*). Weights and information on lactation included. Natural history, Parasites, Wildlife management. Cervidae.

988 **Shaw, M.** 1986. Big game disease research. Okla. Dep. Wildl. Cons. Job 01, Study 03, W-80-R-25. 3 pp.—Limited summary data on diseases and parasites of white-tailed deer (*Odocoileus virginianus*) based on sampling of 28 does. Weights and information on lactation included. Parasites, Wildlife management. Cervidae.

989 **Shaw, M., and R. Umber.** 1979. Big game disease research. Okla. Dep. Wildl. Cons. Job 02, Study 03, W-80-R-18. 7 pp.—Baseline data on diseases and parasites of white-tailed deer (*Odocoileus virginianus*). Natural history, Parasites, Wildlife management. Cervidae.

990 **Shaw, M., and R. Umber.** 1983. Oklahoma white tails. Outdoor Okla. 39(6):2-7.—Popular article on past, present, and future of Oklahoma white-tailed deer (*Odocoileus virginianus*). History, Distribution, Ecology, Populations, Economics, Wildlife management, Popular. Cervidae.

991 **Shaw, M., R. Umber, D. Hoar, and R. Masters.** 1983. Big game season harvest data. Okla. Dep. Wildl. Cons. Job 05, Study 01, W-80-R-22. 55 pp.—Annual harvest data for white-tailed deer (*Odocoileus virginianus*) and elk (*Cervus elaphus*). Information from deer check stations on age, sex, and location. Includes Department's 1982 big game report. Distribution, Natural history, Economics, Wildlife management. Cervidae.

992 **Shaw, M., R. Umber, and R. Masters.** 1984. Big game season harvest data. Okla. Dep. Wildl. Cons. Job 05, Study 01, W-80-R-23. 26 pp.—Annual harvest data for white-tailed deer (*Odocoileus virginianus*) and elk (*Cervus elaphus*). Information from deer check stations on age, sex, and location. Summary maps included. Distribution, Natural history, Economics, Wildlife management. Cervidae.

993 **Shaw, M., R. Umber, R. Thackston, and R. Hoar.** 1982. Big game season harvest data. Okla. Dep. Wildl. Cons. Job 01, Study 01, W-80-R-21. 64 pp.—Annual harvest data for white-tailed deer (*Odocoileus virginianus*) and elk (*Cervus elaphus*). Information from deer check stations on age, sex, and location. Distribution, Natural history, Economics, Wildlife management. Cervidae.

Shaw, M., *see* 1119, 1120, 681, 682, 683.

994 **Shaw, M. G.** 1980. Big game disease research. Okla. Dep. Wildl. Cons. Job 02, Study 03, W-80-R-19. 16 pp.—Baseline data on diseases and parasites of white-tailed deer (*Odocoileus virginianus*). Natural history, Parasites, Wildlife management. Cervidae.

995 **Shaw, M. G.** 1981. Disease monitoring. Okla. Dep. Wildl. Cons. Job 05, Study 05, W-82-R-20. 3 pp.—Compiled available information on disease conditions in Oklahoma. Furbearers. Parasites, Bibliographies. Mammalia.

996 **Shaw, M. G.** 1982. Big game disease research. Okla. Dep. Wildl. Cons. Job 01, STudy 03, W-80-R-21. 43 pp.—Baseline data on diseases and parasites of white-tailed deer (*Odocoileus virginianus*). Includes copy of general booklet to be published by M. G. Shaw and A. A. Kocan. Natural history, Parasites, Wildlife management. Cervidae.

997 **Shaw, M. G.** 1982. Disease monitoring. Okla. Dep. Wildl. Cons. Job 05, Study 05, W-82-R-21. 2 pp.—Reports of rabies in skunks, bats, foxes, raccoons, coyotes, and opossums for 1981, with totals for 1971-81. Parasites, Wildlife management. Didelphidae, Vespertilionidae, Molossidae, Canidae, Procyonidae, Mustelidae.

998 **Shaw, M. G., and A. A. Kocan.** 1981. The parasites of white-tailed deer in Oklahoma. Oklahoma Department of Wildlife Conservation, Oklahoma City. 23 pp.—Provides Oklahoma distribution maps for a number of parasites found in white-tailed deer (*Odocoileus virginianus*). Distribution, Parasites, Wildlife management. Cervidae.

Shellhammer, H. S., *see* 603.

Shepard, W. D., *see* 179.

Shrader, C. A., *see* 505.

999 **Shull, A. M.** 1985. Age-specific reproductive behavior of male American bison. M.S. Thesis, Oklahoma State Univ., Stillwater. 77 pp.—*Bison bison.* Natural history, Behavior. Bovidae.

Shump, A. U., *see* 363, 1000, 1001.

1000 **Shump, K. A., Jr., and A. U. Shump.** 1982. *Lasiurus borealis.* Mamm. Species 183:1-6.—*Lasiurus b. borealis* distribution includes all of Oklahoma. Summary of information in literature. Systematics, Paleontology, Distribution, Natural history, Behavior, Ecology, Parasites, Genetics, Morphology, Physiology, Bibliographies. Vespertilionidae.

1001 **Shump, K. A., Jr., and A. U. Shump.** 1982. *Lasiurus cinereus.* Mamm. Species 185:1-5.—Distribution of *Lasiurus cinereus* includes all of Oklahoma. Summary of information in literature. Systematics, Paleontology, Distribution, Natural history, Behavior, Ecology, Populations, Parasites, Genetics, Morphology, Physiology, Bibliographies. Vespertilionidae.

Shump, K. A., Jr., *see* 362, 363.

Silker, T. H., *see* 788.

Simpson, C. D., *see* 453.

1002 **Sisk, M. E., and C. J. McCoy.** 1963. Stomach contents of *Natrix r. rhombifera* (Reptilia: Serpentes) from an Oklahoma lake. Proc. Okla. Acad. Sci. 44:68-71.—Stomach contents included hispid cotton rats (*Sigmodon hispidus*). Natural history. Cricetidae.

1003 **Skinner, M. P.** 1922. The prong-horn. J. Mammal. 3:82-105.—Review of information on *Antilocapra americana. Antilocapra a. americana* in Oklahoma. Systematics, Distribution, Natural history. Antilocapridae.

Skryja, D. D., *see* 253.

1004 **Slaughter, B. H.** 1961. The significance of *Dasypus bellus* (Simpson) in Pleistocene local faunas. Tex. J. Sci. 13:311-315.—Includes mention of *Dasypus bellus* from Harper County. Paleontology, Ecology. Dasypodidae.

1005 **Slaughter, B. H.** 1967. The vertebrates of the Domebo local fauna, Pleistocene of Oklahoma. Contr. Mus. Great Plains 1:31-35.—Discussion includes seven mammal species found in association with mammoth kill site in Caddo County. Paleontology, Distribution. Geomyidae, Heteromyidae, Cricetidae, Bovidae, Elephantidae.

1006 **Smith, C. C.** 1940. The effect of overgrazing and erosion upon the biota of the mixed-grassed prairie of Oklahoma. Ecology 21:381-397. —Cleveland, Logan, McClain and Oklahoma counties. Includes changes in flora and fauna. Mammals more or less abundant depending on whether degree of grazing brings about habitat conditions nearer or farther from their optimum. Ecology. Mammalia.

1007 **Smith, C. C.** 1940. Notes on the food and parasites of the rabbits of a lowland area in Oklahoma. J. Wildl. Manage. 4:429-431.—*Sylvilagus floridanus* and *S. aquaticus* studied in Cleveland County. Seven species of helminth parasites found. Natural history, Ecology, Parasites. Leporidae.

Smith, C. C., *see* 107.

1008 **Smith, C. F.** 1954. Four new species of cestodes of rodents from the High Plains, central and southern Rockies and notes on *Catenotaenia dendritica.* J. Parasitol. 40:245-254.—Specimens from Oklahoma, Kansas, New Mexico, Colorado, Wyoming and South Dakota. Descriptions of *Hymenandrya thomomyis* (Anoplocephalidae), *Raillietina* (Raillietina), *Sigmodontis* (Davaineidae), *Catenotaenia peromysci* (Catenotainiidae) and *Catenotaenia laguri. Raillietina (r.) sigmodontis* type locality in Alfalfa County, Oklahoma. Parasites. Geomyidae, Cricetidae.

1009 **Smith, H. M.** 1934. Notes on some bat-flies of southern Kansas and northern Oklahoma. J. Kans. Entomol. Soc. 7:62-64.—*Trichobius quadrisetosus* abundant on *Corynorhinus rafinesquii* (=*Plecotus townsendii*) *pallescens,* but not on *Tadarida mexicana* (=*brasiliensis*), in Woods County. Distribution, Parasites. Vespertilionidae, Molossidae.

1010 **Smith, J. D.** 1964. Second record of the eastern harvest mouse from Oklahoma. Trans. Kans. Acad. Sci. 67:204-205.—Specimens from Bryan County assigned to *Reithrodontomys humulis merriami.* Systematics, Distribution. Cricetidae.

Smith, J. F., *see* 213.

1011 **Smith, L.** 1979. Strays and accidentals. Outdoor Okla. 35(3):10-13. —Discusses recent occurrences of *Felis concolor, Ursus americanus, Lutra canadensis,* and other uncommon mammals in Oklahoma. History, Distribution, Popular. Erethizontidae, Canidae, Ursidae, Procyonidae, Mustelidae, Felidae.

Smith, M. H., *see* 88.

1012 **Smith, R.** 1982. Antelope monitoring. Okla. Dep. Wildl. Cons. Job 02, Study 01, W-80-R-21. 5 pp.—Aerial surveys of pronghorns (*Antilocapra americana*) flown in August 1981 and March 1982 in Cimarron County, as well as March in Texas County. Total Oklahoma pronghorns in winter estimated to be 238. Distribution, Populations, Wildlife management. Antilocapridae.

1013 **Smith, R.** 1983. Antelope monitoring. Okla. Dep. Wildl. Cons. Job 02, Study 01, W-80-R-22. 3 pp.—March 1983 aerial survey of pronghorn (*Antilocapra americana*) flown in Cimarron County produced estimate of 370 animals for county. Distribution, Populations, Wildlife management. Antilocapridae.

1014 **Smith, R.** 1984. Antelope monitoring. Okla. Dep. Wildl. Cons. Job 02, Study 01, W-80-R-23. 4 pp.—Late-winter aerial surveys of pronghorn (*Antilocapra americana*) flown in Texas County (1980-82) and Cimarron County (1980-84). Population estimates based on a 60% observation rate provided for 1970-84. Distribution, Populations, Wildlife management. Antilocapridae.

Smith, R., *see* 1119, 1120.

1015 **Smolen, M. J.** 1981. *Microtus pinetorum.* Mamm. Species 147:1-7.—*Microtus pinetorum nemoralis* in eastern half of Oklahoma. Summary of information in literature. Systematics, Paleontology, Distribution, Natural history, Behavior, Ecology, Populations, Parasites, Genetics, Morphology, Physiology. Cricetidae.

Smythe, N., *see* 171.

1016 **Snyder, D. P.** 1982. *Tamias striatus.* Mamm. Species 168:1-8.—Summary of published information on *Tamias striatus. Tamias s. venustus* occurs in eastern Oklahoma. Systematics, Paleontology, Distribution, Natural history, Behavior, Ecology, Parasites, Genetics, Morphology, Bibliographies. Sciuridae.

Sowls, L. K., *see* 707.

1017 **Spencer, D. L.** 1968. Sympatry and hybridization of the eastern and southern plains wood rats. Ph.D. Dissertation, Oklahoma State Univ., Stillwater. 85 pp.—Assessment of habits and habitats of sympatric and hybridizing *Neotoma floridana* and *N. micropus.* Distribution, Natural history, Behavior, Ecology, Genetics, Morphology. Cricetidae.

Spencer, D. L., *see* 101, 707.

1018 **Spencer, S. R., and G. N. Cameron.** 1982. *Reithrodontomys fulvescens.* Mamm. Species 174:1-7.—Summary of published information on *Reithrodontomys fulvescens.* Systematics, Paleontology, Distribution, Natural history, Behavior, Ecology, Genetics, Morphology, Bibliographies. Cricetidae.

Spencer, S. R., *see* 216.

1019 **Sperry, C. C.** 1934. Winter food habits of coyotes: A report of progress, 1933. J. Mammal. 15:286-290.—*Canis latrans* stomachs examined from 10 western states. Summary of food items and illustration of percentage in diet (by bulk). Natural history, Ecology. Canidae, Mammalia.

Sperry, C. C., *see* 782.

1020 **Sprague, J. M.** 1938. Notes on hibernating bats. J. Mammal. 19:500.—*Myotis velifer* and *Plecotus townsendii* in Blaine County caves. Hair of both species in stomachs of both species. Behavior, Parasites. Vespertilionidae.

1021 **Stangl, F. B., Jr.** 1984. Dynamics of chromosomal variation between chromosomally characterized races of *Peromyscus leucopus* (Rodentia: Cricetidae). Ph.D. Dissertation, Texas Tech Univ., Lubbock. 55 pp. —Karyotypes of white-footed mice analyzed in zone of contact. Systematics, Distribution, Genetics. Cricetidae.

1022 **Stangl, F. B., Jr.** 1986. Aspects of a contact zone between two chromosomal races of *Peromyscus leucopus* (Rodentia: Cricetidae). J. Mammal. 67:465-473.—Karyotypes of white-footed mice analyzed along east-west transect in central Oklahoma. Zone of contact between two chromosomal races documented. Systematics, Distribution, Genetics. Cricetidae.

1023 **Stangl, F. B., Jr., and R. J. Baker.** 1984. A chromosomal subdivision in *Peromyscus leucopus:* Implications for the subspecies concept as applied to mammals. Festschrift for Walter W. Dalquest in honor of his sixty-sixth birthday (N. V. Horner, ed.). Biology Department, Midwestern State Univ., Wichita Falls, Texas. 163 pp.—*Peromyscus leucopus* exhibits two distinctive cytotypes and hybrid zone in central Oklahoma. Systematics, Genetics. Cricetidae.

1024 **Stangl, F. B., Jr., and W. W. Dalquest.** 1986. Two noteworthy records of Oklahoma mammals. Southwest. Nat. 31:123-124.—Pygmy mouse (*Baiomys taylori*) reported for first time from Oklahoma (Cotton County). Range is expanding. Also, well preserved jaw of post-Pleistocene (and probably pre-Columbian) black bear (*Ursus* *americanus*) found in Tillman County. Paleontology, Distribution. Cricetidae, Ursidae.

Stangl, F. B., Jr., *see* 100.

Stanley, A. J., *see* 697.

1025 **Stanphill, A.** 1955. Wildlife habitat improvement: Habitat improvement project. Okla. Dep. Wildl. Cons. Job 05, W-024-D-10. 81 pp.—Habitat management study includes discussion of *Sylvilagus.* Natural history, Ecology, Wildlife management. Leporidae.

1026 **Stanphill, A.** 1956. Wildlife habitat improvement: Final report—habitat improvement project. Okla. Dep. Wildl. Cons. Fin. Job 05, W-024-D-11. 48 pp.—Study of *Sylvilagus* populations in Cimarron, Texas, and Beaver counties. Natural history, Ecology, Wildlife management. Leporidae.

1027 **Starrett, A.** 1956. Pleistocene mammals from the Berends fauna of Oklahoma. J. Paleontol. 30:1187-1192.—Discussion of 14 species and seven families of fossil mammals from Beaver County. Systematics, Paleontology. Mammalia.

Stebler, A. M., *see* 351.

1028 **Steele, J. L.** 1959. Canton Reservoir game management area: Controlled grazing program. Okla. Dep. Wildl. Cons. Job 23, W-038-D-04. 4 pp.—Study in Blaine County of effects of cattle upon habitat availability for small mammals, including *Sigmodon hispidus.* Natural history, Ecology, Economics, Wildlife management. Cricetidae, Bovidae.

1029 **Steele, J. L.** 1959. Canton Reservoir game management area: Population control—cotton rats. Okla. Dep. Wildl. Cons. Job 16, W-038-D-04. 3 pp.—Discussion includes *Sylvilagus, Dipodomys, Sigmodon hispidus, Mustela vison,* and *Mephitis mephitis.* Natural history, Ecology, Populations, Techniques, Economics, Wildlife management. Leporidae, Heteromyidae, Cricetidae, Mustelidae, Mammalia.

1030 **Steele, J. L.** 1961. Canton Reservoir game management area: Herbaceous seedlings. Okla. Dep. Wildl. Cons. Job 13, W-038-D-05. 2 pp.—Study in Blaine County of habitat requirements for *Sylvilagus* species. Natural history, Ecology, Wildlife management. Leporidae.

1031 **Steele, J. L.** 1961. Canton Reservoir game management area: Investigations of quail populations and the characteristics of their habitat. Okla. Dep. Wildl. Cons. Job 25, W-038-D-05. 31 pp.—Blaine County. Includes mention of *Didelphis marsupialis* (=*virginianus*) and *Procyon* *lotor.* Natural history, Ecology, Economics. Didelphidae, Procyonidae.

1032 **Steele, J. L.** 1961. Canton Reservoir game management area: Rodent population trends. Okla. Dep. Wildl. Cons. Job 22, W-038-D-06. 4 pp.—Blaine County habitat study includes discussion of *Sylvilagus floridanus, Dipodomys, Sigmodon hispidus,* and *Peromyscus maniculatus.* Natural history, Ecology, Populations, Parasites, Morphology, Techniques, Economics, Wildlife management. Leporidae, Heteromyidae, Cricetidae.

1033 **Steele, J. L.** 1966. Investigation of Comanche County deer herd: Deer production analysis. Okla. Dep. Wildl. Cons. Job 04, W-087-R-01. 11 pp.—Study of Comanche County *Odocoileus virginianus* populations indicates predation by *Canis familiaris, C. latrans,* and *Lynx* (=*Felis*) *rufus.* Natural history, Ecology, Populations, Parasites, Morphology, Techniques, Economics, Wildlife management, Bibliographies. Canidae, Felidae, Cervidae.

1034 **Steele, J. L.** 1966. Investigation of Comanche County deer herd: Evaluation of the range. Okla. Dep. Wildl. Cons. Job 03, W-087-R-01. 13 pp.—Evaluation of Comanche County habitat for *Cervus canadensis* (=*elaphus*), *Odocoileus virginianus, Bison bison, Bos taurus,* and *Lepus californicus.* Natural history, Ecology, Populations, Morphology, Techniques, Economics, Wildlife management, Bibliographies. Leporidae, Cervidae, Bovidae.

1035 **Steele, J. L.** 1967. Investigation of Comanche County deer herd: Evaluation of the range and its use. Okla. Dep. Wildl. Cons. Job 03, W-087-R-02. 13 pp.—Study of Comanche County habitat and populations of *Cervus canadensis* (=*elaphus*) *nelsoni, Odocoileus virginianus,* and *Bison bison.* Natural history, Behavior, Ecology, Populations, Morphology, Techniques, Wildlife management. Cervidae, Bovidae.

1036 **Steele, J. L.** 1967. Investigation of Comanche County deer herd: Search of literature and interviews with deer research workers. Okla. Dep. Wildl. Cons. Job 02, W-087-R-02. 7 pp.—Habitat management study for *Odocoileus virginianus* in Comanche County. Wildlife management, Bibliographies. Cervidae.

1037 **Steele, J. L.** 1968. Investigation of Comanche County deer herd. Okla. Dep. Wildl. Cons. Fin. W-087-R-03. 20 pp.—Study of Comanche County populations of *Cervus canadensis* (=*elaphus*) and *Odocoileus virginianus,* including evaluation of predation by *Canis latrans* and *Lynx* (=*Felis*) *rufus.* Natural history, Behavior, Ecology. Canidae, Felidae, Cervidae.

1038 **Steele, J. L.** 1968. Investigation of Comanche County deer herd: Interviews, literature survey and segment report. Okla. Dep. Wildl. Cons. Job 02, W-087-R-03. 6 pp.—Study of Comanche County populations of *Odocoileus virginianus.* Natural history, Ecology, Populations, Parasites. Cervidae.

1039 **Steele, J. L.** 1969. An investigation of the Comanche County deer herd. Okla. Dep. Wildl. Cons. Fin. W-087-R. 20 pp.—*Odocoileus virginianus* populations in Comanche County studied. Populations. Cervidae.

1040 **Stephens, J. J.** 1960. Stratigraphy and paleontology of a Late Pleistocene Basin, Harper County, Oklahoma. Geol. Soc. Am. Bull. 71:1675-1702.—Lists 20 species and 12 families of mammals from Doby Springs local fauna, Tillman County. Paleontology. Mammalia.

Stobaugh, D. G., *see* 789.

1041 **Stock, A. D.** 1971. Chromosome evolution in the genus *Dipodomys* and its phylogenetic implications. Ph.D. Dissertation, Univ. Oklahoma, Norman. 73 pp.—Karyotypic study of genus *Dipodomys.* Some Oklahoma localities. Systematics, Genetics. Heteromyidae.

1042 **Stock, A. D.** 1974. Chromosome evolution in the genus *Dipodomys* and its taxonomic and phylogenetic implications. J. Mammal. 55:505-526.—*Dipodomys* karyotypic study includes Oklahoma localities. Systematics, Genetics. Heteromyidae.

Stone, A. J., *see* 927.

1043 **Storrs, E. E., and B. Labadies.** 1982. The astonishing armadillo. Natl. Geogr. Mag., 161:820-830.—Includes distribution map showing timing of range expansion of nine-banded armadillos (*Dasypus novemcinctus*) into Oklahoma. History, Distribution, Natural history, Behavior, Ecology, Populations, Popular. Dasypodidae.

1044 **Stout, G. G.** 1971. Nationwide survey of antlerless harvest as a deer management tool. Okla. Dep. Wildl. Cons. Fin. W-109-R-01. 32 pp. —Study of population management of *Odocoileus hemionus* and *O. virginianus.* Populations, Morphology, Wildlife management. Cervidae.

1045 **Stout, G. G.** 1982. Effects of coyote reduction on white-tailed deer productivity on Fort Sill, Oklahoma. Wildl. Soc. Bull. 10:329-332.—Removal of *Canis latrans* from study area increased *Odocoileus virginianus* fawn production. Natural history, Populations, Wildlife management. Canidae, Cervidae.

1046 **Stout, G. G., F. C. Lowry, and F. Carlile.** 1973. The status of elk transplants in eastern Oklahoma. Proc. Annu. Conf. Southeast. Assoc. Game Fish Comm. 26:202-203.—*Cervus elaphus* from Comanche County released in northeastern and southeastern Oklahoma. Distribution, Populations, Parasites, Wildlife management. Cervidae.

1047 **Stovall, J. W.** 1936. A Recent grizzly-bear skull found in Oklahoma. Am. Midl. Nat. 17:781-783.—Skull of *Ursus horribilis nelsoni.* Paleontology, Distribution, Morphology. Ursidae.

1048 **Stovall, J. W.** 1937. *Euceratherium bizzelli,* a new ungulate from Oklahoma. J. Paleontol. 11:450-455.—Description of *Euceratherium bizzelli* from Caddo County. Systematics, Paleontology. Bovidae.

1049 **Stovall, J. W.** 1940. *Megalonyx hogani,* a new species of ground sloth from Gould, Oklahoma. Am. J. Sci. 238:140-146.—Paleontology, Morphology. Megalonychidae.

1050 **Stovall, J. W., and C. W. Hibbard.** 1948. The giant armadillo, *Holmesina septentrionalis* (Leidy), new to the Pleistocene of Oklahoma. J. Mammal. 29:420.—*Holmesina septentrionalis* in Caddo County. Paleontology. Dasypodidae.

1051 **Stovall, J. W., and C. S. Johnston.** 1934. Hypertrophy in the jaw of an Oklahoma proboscidean. Am. Midl. Nat. 15:622-624.—Hypertrophied jaw of *Elephas imperator.* Paleontology, Morphology. Elephantidae.

1052 **Stovall, J. W., and C. S. Johnston.** 1934. *Tapirus haysii* of Oklahoma. Am. Midl. Nat. 15:92-93.—From Pleistocene. Paleontology, Morphology. Tapiridae.

1053 **Stovall, J. W., and C. S. Johnston.** 1935. Two fossil grizzly bears from the Pleistocene of Oklahoma. J. Geol. 43:208-213.—Subspecies *Ursus horribilis oklahomensis* based on these Pleistocene specimens. Paleontology, Distribution, Morphology. Ursidae.

1054 **Stovall, J. W., and J. T. Self.** 1936. A new specimen of *Symbos* from Chickasha, Oklahoma. J. Mammal. 17:422.—*Symbos cavifrons* from Grady County. Paleontology. Bovidae.

1055 **Stovall, J. W., and J. T. Self.** 1936. A new specimen of *Symbos* from Chickasha, Oklahoma. Proc. Okla. Acad. Sci. 16:74.—Specimen of *Symbos* from Grady County. Paleontology. Bovidae.

1056 **Stovall, J. W., and W. S. Strain.** 1936. A new Tertiary mammal deposit in western Oklahoma. Proc. Okla. Acad. Sci. 16:72.—Ellis County locality. Paleontology. Mammalia.

Strain, W. S., *see* 1056.

1057 **Strandtmann, R. W., and O. E. Hunt.** 1951. Two new species of Macronyssidae, with notes on some established genera (Acarina). J. Parasitol. 37:460-470.—*Hirstionyssus staffordi* collected from *Mephitis elongata* (=*mephitis*) in Oklahoma. Parasites. Mustelidae.

1058 **Streubel, D. P., and J. P. Fitzgerald.** 1978. *Spermophilus spilosoma.* Mamm. Species 101:1-4.—*Spermophilus spilosoma* distribution includes Oklahoma Panhandle. Summary of information in literature. Systematics, Paleontology, Distribution, Natural history, Behavior, Ecology, Populations, Parasites, Genetics, Morphology, Physiology, Bibliographies. Sciuridae.

1059 **Streubel, D. P., and J. P. Fitzgerald.** 1978. *Spermophilus tridecemlineatus.* Mamm. Species 103:1-5.—*Spermophilus tridecemlineatus* distribution includes Oklahoma. Summary of information in literature. Systematics, Paleontology, Distribution, Natural history, Behavior, Ecology, Populations, Parasites, Morphology, Physiology, Bibliographies. Sciuridae.

1060 **Stromp, B.** 1984. The bobcat . . . Oklahoma's secretive feline predator. Outdoor Okla. 40(1):30-35.—Popular article on *Felis rufus* biology. History, Distribution, Natural history, Behavior, Ecology, Populations, Economics, Wildlife management, Popular. Felidae.

1061 **Stromp, B.** 1985. The beaver in Oklahoma. Outdoor Okla. 41(1):14-21.—*Castor canadensis.* History, Natural history, Ecology, Populations, Popular. Castoridae.

1062 **Strong, W. E.** 1960. Canadian River hunt. Univ. Oklahoma Press, Norman. 47 pp.—Reproduction of journal covering an 1878 hunting trip in Indian Territory. Localities well described. Mentions mammals encountered. History, Distribution, Popular. Mammalia.

1063 **Sturgis, R. S.** 1939. The Wichita Mountains Wildlife Refuge. Chicago Nat. 2:9-20.—Includes a list of 32 species of mammals from Comanche County refuge. History, Distribution, Wildlife management. Mammalia.

Suttkus, R. D., *see* 226.

1064 **Sutton, G. M.** 1985. Eastern screech-owl eats red bat. Bull. Okla. Ornithol. Soc. 18:22-23.—Red bat (*Lasiurus borealis*) eaten by eastern screech-owl (*Otus asio*). Distribution, Ecology. Vespertilionidae.

Svec, S., *see* 876.

1065 **Svoboda, P. L., J. R. Choate, and R. K. Chesser.** 1985. Genetic relationships among southwestern populations of the Brazilian free-tailed bat. J. Mammal. 66:444-450.—Electrophoretic analysis used to evaluate whether populations of *Tadarida brasiliensis* with different migratory pathways are genetically distinct. Three Oklahoma localities sampled. Systematics, Distribution, Genetics. Molossidae.

Swenson, F. E., *see* 178.

Swift, V. M., *see* 450.

1066 **Sykora, J. F.** 1956. Antelope investigation: Movements of antelope after release. Okla. Dep. Wildl. Cons. Job 01, W-070-R-01. 3 pp.—Study of transplant population of *Antilocapra americana* in Osage County. Distribution, Natural history, Behavior, Ecology, Morphology, Techniques, Wildlife management. Antilocapridae.

1067 **Sykora, J. F.** 1957. Antelope investigation: Success of reproduction. Okla. Dep. Wildl. Cons. Job 02, W-070-R-02. 3 pp.—Study of transplant population of *Antilocapra americana* in Osage County. Mentions *Lepus californicus* and *Canis latrans.* Natural history, Ecology, Populations, Morphology, Wildlife management. Leporidae, Canidae, Antilocapridae.

1068 **Sykora, J. F.** 1959. Antelope investigation: Movements and behavior of antelope. Okla. Dep. Wildl. Cons. Job 01, W-070-R-02. 4 pp.—Study of transplant population of *Antilocapra americana* in Osage County. Distribution, Natural history, Ecology, Morphology, Techniques, Economics, Wildlife management. Antilocapridae.

1069 **Taber, F. W.** 1939. Extension of the range of the armadillo. J. Mammal. 20:489-493.—*Dasypus novemcinctus* in Carter and Creek counties. Distribution. Dasypodidae.

1070 **Talmage, R. V., and G. D. Buchanan.** 1954. The armadillo (*Dasypus novemcinctus*). A review of its natural history, ecology, anatomy and reproductive biology. Rice Inst. Pamphlet 41(2):1-135.—Distribution of *Dasypus novemcinctus* includes southern and west-central Oklahoma. Extensive treatise on armadillo biology. Systematics, Paleontology, Distribution, Natural history, Behavior, Ecology, Parasites, Genetics, Morphology, Physiology, Bibliographies. Dasypodidae.

Talmage, R. V., *see* 185.

1071 **Tate, R. C., and L. B. Nice.** 1927. Say's chipmunk—*Eutamias quadrivittatus quadrivittatus* in Cimarron County. Proc. Okla. Acad. Sci.

7:134.—First record of *Eutamias* (=*Tamias*) *q. quadrivittatus* in Oklahoma. Distribution. Sciuridae.

1072 **Taylor, D. W., and C. W. Hibbard.** 1955. A new Pleistocene fauna from Harper County, Oklahoma. Okla. Geol. Surv. Circ. 37:1-23.—Includes description of *Dasypus bellus* materials from Bar M local fauna, Harper County. Paleontology. Dasypodidae.

Taylor, D. W., *see* 576, 577.

1073 **Taylor, J.** 1964. Additional mammals for Bryan County, Oklahoma. J. Mammal. 45:640-642.—Three additional species found in Bryan County. *Peromyscus gossypinus, P. boylei* (=*attwateri*), and *Rattus rattus.* Distribution. Cricetidae, Muridae.

1074 **Taylor, J.** 1964. Noteworthy predation of the guano bat. J. Mammal. 45:300-301.—Predation of *Tadarida brasiliensis* by Mississippi kites (*Ictinia mississippiensis*), a great horned owl (*Bubo virginianus*), and a barn owl (*Tyto alba*) in Greer County. Ecology. Molossidae.

1075 **Taylor, J., and H. McCarley.** 1963. Vertical distribution of *Peromyscus leucopus* and *P. gossypinus* under experimental conditions. Southwest. Nat. 8:107-108.—*Peromyscus leucopus* and *P. gossypinus* from Marshall County, Oklahoma and Nacogdoches County, Texas. Distribution, Natural history, Behavior, Ecology, Populations. Cricetidae.

1076 **Taylor, J. W.** 1986. A canyon in western Canadian County: Archeological and geomorphological clues from non-destructive testing. Bull. Okla. Anthropol. Soc. 35:69-133.—Bone from surface collection included several mammals. Paleontology. Leporidae, Canidae, Cervidae, Bovidae.

1077 **Taylor, R. J.** 1964. The badger (*Taxidea taxus*) from Johnston County, Oklahoma. Proc. Okla. Acad. Sci. 45:80-81.—Records of Oklahoma occurrences of *Taxidea taxus.* Distribution, Ecology. Mustelidae.

1078 **Taylor, W. P.** 1956. The deer of North America. Stackpole Co., Harrisburg, Pennsylvania, and Wildlife Management Institute, Washington, D.C. 668 pp.—*Odocoileus virginianus texanus* occurs in western Oklahoma, *O. v. macrourus* in eastern, and *O. h. hemionus* in Cimarron County. Systematics, Distribution, Natural history, Behavior, Ecology, Morphology, Wildlife management, Popular. Cervidae.

1079 **Thackston, R.** 1983. Impacts of industrial forest management practices on whitetail deer and eastern wild turkey populations. Okla.

Dep. Wildl. Cons. Job 01, Study 05, W-80-R-22. 22 pp.—Data on white-tailed deer (*Odocoileus virginianus*) and coyotes (*Canis latrans*) from spotlight counts, and additional information on deer from helicopter surveys in southeastern Oklahoma. Comments included on effectiveness of helicopter surveys in pine plantations. Distribution, Natural history, Populations, Ecology, Economics, Wildlife management. Canidae, Cervidae.

1080 **Thackston, R.** 1984. Impacts of industrial forest management practices on whitetail deer and eastern wild turkey populations. Okla. Dep. Wildl. Cons. Job 01, Study 05, W-80-R-23. 16 pp.—Data on white-tailed deer (*Odocoileus virginianus*) and coyotes (*Canis latrans*) from spotlight counts, and additional information on deer from helicopter surveys in southeastern Oklahoma. Distribution, Natural history, Populations, Ecology, Techniques, Economics, Wildlife management. Canidae, Cervidae.

Thackston, R., *see* 993, 789.

Thedford, T. R., *see* 681, 1183.

Thomas, H. H., *see* 130.

1081 **Thomas, J. W., and D. E. Toweill, eds.** 1982. Elk of North America: Ecology and management. Stackpole Books, Harrisburg, Pennsylvania. 698 pp.—Covers range of topics on *Cervus elaphus,* including some information on Oklahoma. Systematics, History, Distribution, Natural history, Behavior, Ecology, Populations, Morphology, Wildlife management. Cervidae.

1082 **Thompson, J. N., Jr.** 1969. Variations in the interparietal of the kangaroo rat, *Dipodomys ordii.* Am. Midl. Nat. 82:625-628.—Interparietal highly variable in *Dipodomys ordii.* Genetics, Morphology. Heteromyidae.

1083 **Thompson, J. N., Jr.** 1971. Chromosomes of Oklahoma mammals. I: Three species of squirrels. Proc. Okla. Acad. Sci. 51:79-80.—*Eutamias* (=*Tamias*) *q. quadrivittatus, Tamias striatus venustus,* and *Spermophilus tridecemlineatus texensis.* Karyology. Delaware, Cimarron, and Cleveland counties. Genetics. Sciuridae.

1084 **Thompson, J. N., Jr., and S. D. Barrett.** 1967. A nest complex of *Perognathus hispidus* (Rodentia: Heteromyidae) in Oklahoma. Proc. Okla. Acad. Sci. 48:105-108.—*Perognathus* (=*Chaetodipus*) *hispidus spilotus.* Nest structure described. Natural history, Ecology. Heteromyidae.

1085 **Thompson, J. N., Jr., and J. K. Greer.** 1967. Oklahoma range of *Dipodomys ordii* (Rodentia: Heteromyidae). Proc. Okla. Acad. Sci. 48:108-109.—Known range of *Dipodomys ordii* extended. Oklahoma distribution map. Distribution, Ecology. Heteromyidae.

1086 **Thompson, J. N., Jr., and J. K. Greer.** 1970. Geographic ranges of three Oklahoma mammals. Southwest. Nat. 15:131.—Range extensions reported for *Nycticeius humeralis, Dipodomys ordii,* and *Oryzomys palustris.* Distribution. Vespertilionidae, Heteromyidae, Cricetidae.

1087 **Thomson, C. E.** 1982. *Myotis sodalis.* Mamm. Species 163:1-5.—Summary of published information on *Myotis sodalis. Myotis sodalis* occurs in eastern Oklahoma. Systematics, Distribution, Natural history, Behavior, Ecology, Populations, Genetics, Morphology, Bibliographies. Vespertilionidae.

1088 **Timm, R. M., and R. D. Price.** 1980. The taxonomy of *Geomydoecus* (Mallophaga: Trichodectidae) from the *Geomys bursarius* complex (Rodentia: Geomyidae). J. Med. Entomol. 17:126-145.—Study includes lice from Oklahoma *Geomys bursarius* specimens. Distribution, Parasites. Geomyidae.

Timm, R. M., *see* 555.

Toldan, R., Jr., *see* 958.

1089 **Tong, M. E., Jr.** 1962. Hunting elephants in Oklahoma. Arkansas Amateur 1(5):5-7.—Discussion of mammoth excavations in Caddo County and other areas. Paleontology. Elephantidae.

1090 **Tong, M. E., Jr.** 1967. The Cooperton mammoth. Great Plains Newsl. 3(5):9.—Mentions intent to reexamine Kiowa County *Mammuthus* site. Paleontology, Popular. Elephantidae.

1091 **Tong, M. E., Jr.** 1968. Cooperton site dated. Okla. Anthropol. Soc. Newsl. 16(8):3-4.—Three radiocarbon dates on bones of Cooperton mammoth (Kiowa County). Paleontology. Elephantidae.

Torre, L. de la, *see* 589.

Toweill, D. E., *see* 1081.

1092 **Trail, M. A., and R. Tumlison.** 1984. Anomalies of bobcat skulls from Oklahoma. Proc. Okla. Acad. Sci. 64:46-47.—Supernumerary bones, atypically shaped bones, and dental irregularities found in *Felis rufus* skulls. Morphology. Felidae.

Traut, G. L., *see* 358.

1093 **Trowbridge, A. H., and H. L. Whitaker.** 1940. A new kangaroo rat from Oklahoma. J. Mammal. 21:343-345.—*Dipodomys oklahomae* (=*ordii oklahomae*), new species. Systematics, Distribution. Heteromyidae.

1094 **Trowbridge, H. M.** 1938. Analysis of Spiro Mound textiles. Am. Antiquity 4:51-53.—Textiles from the Spiro Mounds composed of hair of rabbits and possibly other mammals. Paleontology, History. Leporidae, Mammalia.

1095 **Tumlison, R., and S. Shalaway.** 1986. An annotated bibliography on the North American river otter, *Lutra canadensis*. Okla. Coop. Fish and Wildlife Res. Unit, Oklahoma State Univ., Stillwater.—Based on extensive search of abstracting sources. Also includes some project reports by furbearer biologists. Bibliographies. Mustelidae.

Tumlison, R., *see* 1092.

1096 **Twente, J. W., Jr.** 1954. Habitat selection of cavern-dwelling bats as illustrated by four vespertilionids. Ph.D. Dissertation, Univ. Michigan, Ann Arbor. 163 pp.—Banding studies of *Myotis velifer, Corynorhinus rafinesquei* (=*Plecotus townsendii*), *Antrozous bunkeri* (=*pallidus*) and *Eptesicus fuscus* in Woods County and adjacent Kansas. Distribution, Natural history, Behavior, Ecology, Populations, Morphology, Physiology, Techniques. Vespertilionidae.

1097 **Twente, J. W., Jr.** 1954. Predation on bats by hawks and owls. Wilson Bull. 66:135-136.—In Woods County, *Tyto alba* and *Bubo virginianus* prey on *Tadarida mexicana* (=*brasiliensis*). Also, *Falco sparverius* preys on *Myotis velifer*. Natural history, Behavior. Vespertilionidae, Molossidae.

1098 **Twente, J. W., Jr.** 1955. Aspects of a population study of cavern-dwelling bats. J. Mammal. 36:379-389.—*Myotis velifer incautus, Corynorhinus rafinesquei* (=*Plecotus townsendii*) *pallescens, Antrozous bunkeri* (=*pallidus*), and *Eptesicus fuscus*. Distribution, Natural history, Ecology, Populations. Vespertilionidae.

1099 **Twente, J. W., Jr.** 1955. Some aspects of habitat selection and other behavior of cavern-dwelling bats. Ecology 36:706-732.—Study area includes four caves in Woods County. Bats studied include *Myotis velifer incautus, Corynorhinus rafinesquei* (=*Plecotus townsendii*) *pallescens, Eptesicus fuscus pallidus,* and *Antrozous bunkeri* (=*pallidus*). Distribution, Natural history, Behavior, Ecology, Physiology. Vespertilionidae.

1100 **Twente, J. W., Jr.** 1956. Ecological observations on a colony of *Tadarida mexicana.* J. Mammal. 37:42-47.—*Tadarida mexicana* (=*brasiliensis*) in Woods County. Study of population size, weight fluctuation, habitat selection, and sex ratio. Distribution, Natural history, Behavior, Ecology, Populations, Physiology. Molossidae.

1101 **Tyler, J. D.** 1965. Habitat types used by the bobwhite quail and other wildlife in southwestern Oklahoma. M.S. Thesis, Oklahoma State Univ., Stillwater. 50 pp.—Study in Kiowa, Comanche, and Tillman counties assessed habitat utilization by 25 species of mammals. Distribution, Ecology. Mammalia.

1102 **Tyler, J. D.** 1968. Distribution and vertebrate associates of the black-tailed prairie dog in Oklahoma. Ph.D. Dissertation, Univ. Oklahoma, Norman. 85 pp.—*Cynomys ludovicianus.* Distribution maps included. History, Distribution, Natural history, Ecology, Populations, Morphology, Wildlife management. Sciuridae, Mammalia.

1103 **Tyler, J. D.** 1970. Vertebrates in a prairie dog town. Proc. Okla. Acad. Sci. 50:110-113.—*Didelphis marsupialis* (=*virginiana*), *Sylvilagus auduboni* (=*audubonii*), *Lepus californicus, Cynomys ludovicianus, Citellus* (=*Spermophilus*) *tridecemlineatus, Peromyscus maniculatus, Felis domesticus* (=*catus*), *Taxidea taxus, Neotoma micropus,* and *Onychomys leucogaster.* Jackson County. Ecology aspects of colonies attract many other animals. Ecology, Economics. Sciuridae, Mammalia.

1104 **Tyler, J. D.** 1977. Dead Cooper's hawk found in packrat nest. Bull. Okla. Ornithol. Soc. 10:32.—Dead *Accipiter cooperi* found in nest of *Neotoma micropus* in Tillman County. Natural history. Cricetidae.

1105 **Tyler, J. D.** 1979. Occurrence of the red fox (*Vulpes vulpes*) in western Oklahoma. Proc. Okla. Acad. Sci. 59:124-125.—*Vulpes vulpes* recorded from four western counties. Distribution. Canidae.

1106 **Tyler, J. D.** 1983. Notes on burrowing owl (*Athene cunicularia*) food habits in Oklahoma. Southwest. Nat. 28:100-102.—Burrowing owl pellets from western Oklahoma *Cynomys ludovicianus* colonies included eight species of mammals. Natural history, Ecology. Soricidae, Sciuridae, Geomyidae, Heteromyidae, Cricetidae.

1107 **Tyler, J. D., and H. N. Buscher.** 1975. Parasites of vertebrates inhabiting prairie dog towns in Oklahoma. I: Ectoparasites. Proc. Okla. Acad. Sci. 55:166-168.—*Cynomys ludovicianus, Spermophilus tridecemlineatus,* and *Sylvilagus audubonii.* Three species of fleas, two species of mallophaga, one species of tick, and one dipteran larva. Vectors present in Oklahoma for an epizootic. Ecology, Parasites. Leporidae, Sciuridae.

1108 **Tyler, J. D., and A. R. Gilliland.** 1979. Status of *Notiosorex crawfordi* in Oklahoma, and new distributional records. Southwest. Nat. 24:375-376.—*Notiosorex crawfordi* skeletal remains found in owl pellets from Jackson, Comanche, and Greer counties. Distribution, Natural history. Soricidae.

1109 **Tyler, J. D., and J. F. Jensen.** 1981. Notes on foods of great horned owls (*Bubo virginianus*) in Jackson County, Oklahoma. Proc. Okla. Acad. Sci. 61:28-30.—*Bubo virginianus* preys on at least nine species of mammals in Jackson County. Natural history, Ecology. Soricidae, Molossidae, Leporidae, Heteromyidae, Cricetidae, Muridae.

1110 **Tyler, J. D., and M. J. Lodes.** 1980. Distributional status of the eastern spotted skunk, *Spilogale putorius interrupta*, in Oklahoma. Proc. Okla. Acad. Sci. 60:102-104.—*Spilogale putorius* occurs throughout state. Distribution. Mustelidae.

1111 **Tyler, J. D., and L. Payne.** 1982. Second Oklahoma record for the silver-haired bat, *Lasionycteris noctivagans*. Southwest. Nat. 27:245.—*Lasionycteris noctivagans* captured in December in Comanche County. Distribution, Natural history. Vespertilionidae.

1112 **Tyler, J. D., and C. M. Scott.** 1982. Status and distribution of *Lasiurus cinereus* (Chiroptera: Vespertilionidae) in Oklahoma. Proc. Okla. Acad. Sci. 62:91-92.—Summary of records of hoary bat (*Lasiurus cinereus*) in Oklahoma. History, Distribution. Vespertilionidae.

Tyler, J. D., *see* 197.

1113 **Umber, R.** 1978. Census techniques evaluation. Okla. Dep. Wildl. Cons. Job 01, Study 03, W-82-R-16. 6 pp.—Updates rabbit (*Sylvilagus* spp.) census data. Populations, Techniques. Leporidae.

1114 **Umber, R.** 1979. Game techniques evaluation. Okla. Dep. Wildl. Cons. Job 01, Study 03, W-82-R-18. 2 pp.—Data on morning and evening rabbit (*Sylvilagus floridanus*) counts. Populations. Leporidae.

1115 **Umber, R.** 1980. Evaluation of summer roadside rabbit counts in Oklahoma. Okla. Dep. Wildl. Cons. Fin. Study 03, Job 01, W-82-R. 18 pp.—Compared morning and evening roadside counts of rabbits (*Sylvilagus* spp.). Populations, Techniques. Leporidae.

1116 **Umber, R.** 1985. Impacts of industrial forest management practices on whitetail deer and eastern wild turkey populations. Okla. Dep.

Wildl. Cons. Job 01, Study 05, W-80-R-24. 2 pp.—Helicopter surveys in southeastern Oklahoma in September to evaluate fawn/doe and buck/doe ratios of white-tailed deer (*Odocoileus virginianus*). Distribution, Natural history, Populations, Economics, Wildlife management. Cervidae.

1117 **Umber, R.** 1986. Impacts of industrial forest management practices on white-tailed deer and eastern wild turkey populations. Okla. Dep. Wildl. Cons. Job 01, Study 05, W-80-R-25. 2 pp.—Helicopter surveys in southeastern Oklahoma in late August and September 1985 to determine fawn/doe and buck/doe ratios of white-tailed deer (*Odocoileus virginianus*). Natural history, Populations, Economics, Wildlife management. Cervidae.

1118 **Umber, R., F. Lowry, F. Carlile, and R. Hoar.** 1979. Investigations of big game species. Okla. Dep. Wildl. Cons. Job 01, Study 01, W-80-R-18. 73 pp.—Includes population and/or harvest data on elk (*Cervus elaphus*), pronghorn (*Antilocapra americana*), and white-tailed deer (*Odocoileus virginianus*). Emphasis on deer. Distribution, Populations, Wildlife management. Cervidae, Antilocapridae.

1119 **Umber, R., M. Shaw, R. Masters, R. Smith, and D. Geary.** 1985. Big game season harvest data. Okla. Dep. Wildl. Cons. Job 04, Study 01, W-80-R-24. 80 pp.—Annual harvest data for white-tailed deer (*Odocoileus virginianus*) and elk (*Cervus elaphus*). Information from deer check stations on age, sex, and location. Summary maps included, as is the Department's 1984 big game report. Natural history, Populations, Economics, Wildlife management. Cervidae.

1120 **Umber, R., M. Shaw, R. Masters, R. Smith, and D. Geary.** 1986. Big game season harvest data. Okla. Dep. Wildl. Cons. Job 04, Study 01, W-80-R-25. 62 pp.—Annual harvest data for white-tailed deer (*Odocoileus virginianus*) and elk (*Cervus elaphus*). Information from deer check stations on age, sex, and location. Summary maps included. Distribution, Natural history, Economics, Wildlife management. Cervidae.

Umber, R., *see* 307, 989, 990, 991, 992, 993.

1121 **Umber, R. W.** 1977. Census techniques evaluation. Okla. Dep. Wildl. Cons. Job 01, Study 03, W-82-R-16. 7 pp.—Roadside rabbit (*Sylvilagus* spp.) counts. Populations, Techniques, Wildlife management. Leporidae.

Uno, G. E., *see* 264.

1122 **Urban, E. K., and R. B. Wimmer.** 1959. *Reithrodontomys megalotis* and *Urocyon cinereoargenteus* from western Oklahoma. J. Mammal. 40:450.—*Reithrodontomys megalotis* and *Urocyon cinereoargenteus* in Texas and Cimarron counties. Distribution. Cricetidae, Canidae.

Urbston, D. F., *see* 354.

1123 **Utsler, H. W.** 1974. An evaluation of potential criteria for estimating age of coyotes. M.S. Thesis, Oklahoma State Univ., Stillwater. 28 pp. —Three methods of aging *Canis latrans* assessed and found to be imprecise. Natural history, Morphology, Techniques. Canidae.

1124 **Valastro, S., Jr., E. M. Davis, and A. G. Varela.** 1978. University of Texas at Austin radiocarbon dates XII. Radiocarbon 20:245-273.—Includes date of 703 ± 190 for Perry Ranch *Bison antiquus* from southwestern Oklahoma. Paleontology. Bovidae.

Van Dyke, T. S., *see* 927.

1125 **Van Gelder, R. G.** 1959. A taxonomic revision of the spotted skunks (genus *Spilogale*). Bull. Am. Mus. Nat. Hist. 117:229-329.—*Spilogale putorius gracilis* occurred in Cimarron County, with *S. p. interrupta* in rest of state. Systematics, Distribution, Morphology. Mustelidae.

1126 **Van Zyll de Jong, C. G.** 1972. A systematic review of the Nearctic and Neotropical river otters (genus *Lutra*, Mustelidae, Carnivora). Royal Ontario Mus., Life Sci. Contrib., No. 80. 104 pp.—Indicates *Lutra canadensis lataxina* was subspecies originally found in Oklahoma. Systematics, History, Distribution, Morphology. Mustelidae.

1127 **Van Zyll de Jong, C. G.** 1984. Taxonomic relationships of Nearctic small-footed bats of the *Myotis leibii* group (Chiroptera: Vespertilionidae). Can. J. Zool. 62:2519-2526.—Specimens of *Myotis ciliolabrum* used from Oklahoma. Morphometric analyses do not support claim that *M. leibii leibii* and *M. leibii melanorhinus* intergrade in Oklahoma. Suggests that western forms of *M. leibii* represent a separate species (*M. ciliolabrum*). Systematics, Distribution, Morphology. Vespertilionidae.

Vance, D., *see* 305.

Varela, A. G., *see* 1124.

1128 **Vaughan, S.** 1975. Archaeological investigations for the Copan Reservoir, northeastern Oklahoma and southeastern Kansas. Okla. River Basin Surv. Archaeol. Site Rep. 29:1-151.—Northern Oklahoma prehistoric sites yielded remains of seven mammal species. Paleontology. Mammalia.

1129 **Vehik, R.** 1982. The archaeology of the Bug Hill site (34Pu-116): Pushmataha County, Oklahoma. Archaeol. Research and Management Center, Univ. Oklahoma. Research Ser. No. 7. 226 pp.—Extensive mammal remains (16 species) found. Habitats of species discussed. Paleontology. Didelphidae, Talpidae, Leporidae, Sciuridae, Geomyidae, Cricetidae, Canidae, Ursidae, Procyonidae, Mustelidae, Cervidae, Bovidae.

1130 **Vehik, R., ed.** 1982. Phase II. Archaeological investigations at Clayton Lake, southeast Oklahoma. Archaeol. Research and Management Center, Univ. Oklahoma. Research Ser. No. 8. 435 pp.—Faunal remains reported for several sites. Relatively few animal remains. White-tailed deer (*Odocoileus virginianus*) identified. Paleontology. Cervidae.

1131 **Vehik, S. C., and P. Flynn.** 1982. Archeological excavations at the early Plains Village Uncas site. Bull. Okla. Anthropol. Soc. 31:5-70.—Faunal remains included white-tailed deer (*Odocoileus virginianus*) and bison (*Bison bison*). Paleontology. Cervidae, Bovidae.

Verts, B. J., *see* 1136.

1132 **Vestal, B. M., and J. J. Hellack.** 1978. Comparison of neighbor recognition in two species of deermice (*Peromyscus*). J. Mammal. 59:339-346.—*Peromyscus leucopus noveboracensis* and *P. maniculatus ozarkiarum.* Differences found between neighbor and stranger interactions. Behavior, Ecology. Cricetidae.

1133 **Vestal, B. M., and G. D. Schnell.** 1986. Influence of environmental complexity and space on social interactions of mice (*Mus musculus* and *Peromyscus leucopus*). J. Comp. Psychol. 100:143-154.—Experimental studies of laboratory animals descended (first to fourth generation) from mice taken in vicinity of Norman, Oklahoma. Behavior. Cricetidae, Muridae.

Vestal, B. M., *see* 823, 1179.

1134 **Villa-R., B.** 1956. *Tadarida brasiliensis mexicana* (Saussure), el murcielago guanero, es una subespecie migratoria. Acta. Zool. Mex. 1(11):1-11.—*Tadarida brasiliensis* banded in northern Mexico recaptured near Freedom, Oklahoma. Distribution, Natural history. Molossidae.

1135 **Wade, O.** 1928. Notes on the time of breeding and the number of young of *Cynomys ludovicianus.* J. Mammal. 9:149-151.—*Cynomys ludovicianus.* Natural history. Sciuridae.

1136 **Wade-Smith, J., and B. J. Verts.** 1982. *Mephitis mephitis.* Mamm. Species 173:1-7.—Summary of published information on *Mephitis mephitis.* Systematics, Paleontology, Distribution, Natural history, Behavior, Ecology, Parasites, Genetics, Morphology, Economics, Bibliographies. Mustelidae.

1137 **Waldrip, G. P.** 1977. Elk habitat use during calving season with possible effects on white-tailed deer at the Wichita Mountains National Wildlife Refuge. M.S. Thesis, Oklahoma State Univ., Stillwater. 81 pp.—Cows and calves of *Cervus canadensis* (=*elaphus*) in Comanche County occupied primarily forested or edge habitats, and appeared to have little interaction with *Odocoileus virginianus.* Natural history, Behavior, Ecology, Techniques, Wildlife management. Cervidae.

Waldrup, K., *see* 681, 683.

1138 **Walker, E. P.** 1975. Mammals of the world (J. L. Paradiso, ed.). Third edition. 3 vols. Johns Hopkins Univ. Press, Baltimore. 1,500 pp.—Discussion of families, genera, and some species. First edition 1964, second 1968, and fourth (which was retitled "Walker's mammals of the world" and edited by R. M. Nowack and J. L. Paradiso). Systematics, History, Distribution, Natural history, Behavior, Ecology, Morphology, Economics, Bibliographies. Mammalia.

1139 **Walker, J. S.** 1975. The effects of crowding upon some adrenal responses in male white-footed mice (*Peromyscus leucopus*). M.S. Thesis, Oklahoma State Univ., Stillwater. 95 pp.—Study of Payne County *Peromyscus leucopus* assesses effects of crowding and isolation on leucocyte count, granuloma formation, adrenal weight, and corticosterone levels. Behavior, Populations, Physiology. Cricetidae.

Wallace, L. L., *see* 886.

1140 **Wallmo, O. C., ed.** 1981. Mule and black-tailed deer of North America. Univ. Nebraska Press, Lincoln. 605 pp.—Wide range of topics covering biology of *Odocoileus hemionus,* which is found in far western Oklahoma. Little information specifically on Oklahoma. Systematics, Distribution, Natural history, Behavior, Ecology, Populations, Parasites, Morphology, Wildlife management. Cervidae.

Wallmo, O. C., *see* 14.

Wample, G. E., *see* 879.

Ward, B. C., *see* 224.

1141 **Ward, C. M.** 1956. Species of the genus *Myotis* in Oklahoma. M.S. Thesis, Oklahoma A and M Coll., Stillwater. 84 pp.—Study assesses status of *Myotis lucifugus, M. yumanensis, M. austroriparius, M. grisescens, M. velifer, M. keeni* (=*keenii*), *M. sodalis,* and *M. subulatus* (=*leibii*). Systematics, Distribution, Morphology. Vespertilionidae.

Ward, C. M., *see* 426.

1142 **Ward, J. W.** 1934. A study of some parasites of rabbits of central Oklahoma. Proc. Okla. Acad. Sci. 14:31-32.—Arthropods, cestodes, nematodes, and protozoans found in *Sylvilagus floridans* (=*floridanus*), *Lepus aquaticus* (=*Sylvilagus aquaticus*), and *Lepus californicus.* Parasites. Leporidae.

Warde, W. D., *see* 924.

Warfel, H. E., *see* 625.

1143 **Warner, J. W.** 1973. The cytogenetics of the plains woodrat (*Neotoma micropus*). Ph.D. Dissertation, Texas Tech Univ., Lubbock. 70 pp.—Study of three cytotypes of *Neotoma micropus* includes specimens from Harmon County. Systematics, Genetics. Cricetidae.

1144 **Warner, J. W.** 1976. Chromosomal variation in the plains woodrat: Geographical distribution of three chromosomal morphs. Evolution 30:593-598.—Chromosomal variation in *Neotoma micropus,* including some from Harmon County. Populations, Genetics, Morphology. Cricetidae.

1145 **Warren, D.** 1974. Oklahoma's endangered species. Outdoor Okla. 30(8):17-19.—*Canis niger* (=*rufus*), *Mustela nigripes,* and *Myotis sodalis.* Account for the layman. Distribution, Ecology, Morphology, Popular. Vespertilionidae, Canidae, Mustelidae.

1146 **Warren, E. R.** 1927. The beaver. Its work and its ways. Am. Soc. Mammal. Monogr. 2:1-177.—Paleontology, History, Distribution, Natural history, Behavior, Parasites. Castoridae.

1147 **Watkins, L. C.** 1972. *Nycticeius humeralis.* Mamm. Species 23:1-4.—*Nycticeius humeralis* found in eastern Oklahoma. Summary of information in literature. Systematics, Distribution, Natural history, Ecology, Parasites, Genetics, Morphology, Bibliographies. Vespertilionidae.

1148 **Watson, V.** 1950. The Optima focus of the Panhandle aspect: Description and analysis. Bull. Tex. Archaeol. Paleontol. Soc. 21:7-68.—Excavation of Texas County site produced bones of several mammal species. Paleontology. Leporidae, Sciuridae, Tayassuidae, Cervidae, Antilocapridae, Bovidae.

1149 **Watts, D. E.** 1970. Burrowing habits of the plains pocket gopher in northcentral and northwestern Oklahoma during the fall and winter of 1969-70. Ed.D. Dissertation, Oklahoma State Univ., Stillwater. 62 pp.—Study of *Geomys bursarius* burrows in Payne and Harper counties. Distribution, Natural history, Behavior, Ecology, Populations, Economics. Geomyidae.

Watts, H. A., *see* 215.

1150 **Webb, S. D.** 1983. A new species of *Pediomeryx* from the Late Miocene of Florida, and its relationships within the subfamily Cranioceratinae (Ruminantia: Dromomerycidae). J. Mammal. 64:261-276.—Analyzed samples of *Pediomeryx,* including *P.* (*Yumaceras*) cf. *figginsi* from Higgins sites, Ogallala Formation and *P.* (*P.*) *hemphillensis* from Guymon, Hemphill beds in Oklahoma. All placed in subfamily Cranioceratinae. Systematics, Paleontology. Dromomerycidae.

1151 **Webster, W. D., and J. K. Jones, Jr.** 1982. *Reithrodontomys megalotis.* Mamm. Species 167:1-5.—Summary of published information on *Reithrodontomys megalotis. Reithrodontomys m. aztecus* occurs in Oklahoma Panhandle. Systematics, Paleontology, Natural history, Behavior, Ecology, Parasites, Genetics, Morphology, Bibliographies. Cricetidae.

1152 **Webster, W. D., and J. K. Jones, Jr.** 1985. Nongeographic variation, reproduction, and demography in the Texas kangaroo rat, *Dipodomys elator* (Rodentia: Heteromyidae). Tex. J. Sci. 37:51-61.—Study of specimens taken in Hardeman County, Texas. Reproduction, demography, sexual dimorphism, and molt evaluated. Natural history, Populations, Morphology, Physiology. Heteromyidae.

Webster, W. D., *see* 226.

1153 **Whitaker, H. L.** 1937. Occurrence of the Texas rice rat in Oklahoma. J. Mammal. 18:102.—*Oryzomys palustris* in McCurtain County. Distribution. Cricetidae.

Whitaker, H. L., *see* 1093.

1154 **Whitaker, J. O., Jr.** 1972. *Zapus hudsonius.* Mamm. Species 11:1-7. —Fossil and Recent distribution of *Zapus hudsonius* includes northern Oklahoma. Summary of information in literature. Systematics, Paleontology, History, Distribution, Natural history, Behavior, Ecology, Parasites, Genetics, Morphology, Physiology, Bibliographies. Zapodidae.

1155 **Whitaker, J. O., Jr.** 1974. *Cryptotis parva.* Mamm. Species 43:1-8.—Distribution of *Cryptotis parva* includes most of Oklahoma. Summary of information in literature. Systematics, Paleontology, History, Distribution, Natural history, Behavior, Ecology, Populations, Parasites, Morphology, Physiology, Bibliographies. Soricidae.

1156 **Whitaker, J. O., Jr.** 1980. The Audubon Society field guide to North American mammals. Alfred A. Knopf, Inc., New York.—Distribution, Behavior, Ecology, Morphology, Techniques, Popular. Mammalia.

1157 **Whitaker, J. O., Jr., and N. Wilson.** 1974. Host and distribution lists of mites (Acari), parasitic and phoretic, in the hair of wild mammals of North America, north of Mexico. Am. Midl. Nat. 91:1-67.—Lists of hosts and mites, by state, compiled from literature sources. Parasites. Mammalia.

Whitaker, J. O., Jr., *see* 864.

1158 **White, J. A.** 1953. Genera and subgenera of chipmunks. Univ. Kans. Publ. Mus. Nat. Hist. 5:543-561.—Genera *Eutamias* (subgenus *Neotamias*) and *Tamias* occur in Oklahoma. Systematics, Distribution, Morphology. Sciuridae.

1159 **White, J. A.** 1953. Taxonomy of the chipmunks, *Eutamias quadrivittatus* and *Eutamias umbrinus.* Univ. Kans. Publ. Mus. Nat. Hist. 5:563-582.—*Eutamias* (=*Tamias*) *q. quadrivittatus* occurs in Cimarron County. Systematics, Distribution, Morphology. Sciuridae.

1160 **Whitenack, D. L., and A. A. Kocan.** 1981. Experimental malignant catarrhal fever (African form) in white-tailed deer. J. Wildl. Dis. 17:443-451.—Parasites. Cervidae.

1161 **Whitenton, R. O., H. W. Orr, R. N. Danielson, and R. C. Hughes.** 1943. Five kittens with conjoined umbilical cords. J. Mammal. 24:270-271.—*Felis catus* kittens of both sexes. Genetics, Morphology. Felidae.

1162 **Whittle, R. K.** 1979. Age in relation to the winter food habits and helminth parasites of the bobcat (*Lynx rufus,* Schreber) in Oklahoma. M.S. Thesis, Oklahoma State Univ., Stillwater. 36 pp.—Young *Lynx* (=*Felis*) *rufus* eat primarily *Sigmodon hispidus,* and adult bobcats eat mostly *Sylvilagus* species. Most or all parasites seem unaffected by age of host. Natural history, Ecology, Parasites. Leporidae, Cricetidae, Felidae.

Wiens, J. A., *see* 913.

Wigtail, G. W., *see* 545.

1163 **Wilcomb, M. J., Jr.** 1954. A study of prairie dog burrow systems and the ecology of their arthropod inhabitants in central Oklahoma. Ph.D. Dissertation, Univ. Oklahoma, Norman. 158 pp.—Study of *Cynomys ludovicianus* colonies in Cleveland and Grady counties. Behavior, Ecology, Parasites. Sciuridae.

1164 **Wilcomb, M. J., Jr., M. E. Griffith, and L. L. Ellis.** 1952. Commensal rat ectoparasite collections in Oklahoma. Public Health Monogr. 5:31-37.—Fleas, lice, and mites collected from *Rattus norvegicus* and *R. rattus* in Oklahoma. Parasites. Muridae.

1165 **Wiley, R. W.** 1980. *Neotoma floridana.* Mamm. Species 139:1-7.—*Neotoma floridana attwateri* in eastern four-fifths of Oklahoma. Summary of information in literature. Systematics, Paleontology, Distribution, Natural history, Behavior, Ecology, Populations, Parasites, Genetics, Morphology, Physiology, Bibliographies. Cricetidae.

1166 **Wilkins, K. T.** 1986. *Reithrodontomys montanus.* Mamm. Species 257:1-5.—Summary of published information on plains harvest mouse, *Reithrodontomys montanus.* Systematics, Paleontology, History, Distribution, Natural history, Behavior, Ecology, Populations, Genetics, Morphology, Bibliographies. Cricetidae.

1167 **Wille, C.** 1976. Controlling nuisance wildlife. Outdoor Okla. 32(2):2-6.—Account for the layman on rats, mice, squirrels, skunks, raccoons, cottontail rabbits, gophers, moles, beavers, bats, prairie dogs, and coyotes. Legal measures for control. Ecology, Economics, Wildlife management, Popular. Mammalia.

1168 **Williams, C. R.** 1954. Mammalian ecology of the Great Salt Plains Wildlife Refuge. Ph.D. Dissertation, Univ. Oklahoma, Norman. 129 pp.—Alfalfa County. Seven mammalian habitat areas. Nineteen species reported. History, Distribution, Natural history, Behavior, Ecology. Mammalia.

1169 **Williams, D. F.** 1978. Systematics and ecogeographic variation of the Apache pocket mouse (Rodentia: Heteromyidae). Bull. Carnegie Mus. Nat. Hist. 10:1-57.—Comparisons made with plains pocket mice (*Perognathus flavescens*) from Oklahoma. Systematics, Distribution. Heteromyidae.

1170 **Williams, G. L.** 1977. Simulation modeling of big game at Wichita Mountains Wildlife Refuge. Ph.D. Dissertation, Colorado State

Univ., Fort Collins. 242 pp.—Population models based on historical data for *Cervus canadensis* (=*elaphus*), *Odocoileus virginianus, Bison bison,* and *Bos taurus* at Wichita Mountains Wildlife Refuge, Comanche County. Checklist of mammals of refuge also included. Natural history, Behavior, Ecology, Populations, Techniques, Wildlife management. Cervidae, Bovidae, Mammalia.

1171 **Williams, G. L.** 1981. An example of simulation models as decision tools in wildlife management. Wildl. Soc. Bull. 9:101-107.—Computer simulation model used to analyze a management decision relative to changes in harvest policy for elk (*Cervus elaphus*) on Wichita Mountains National Wildlife Refuge near Lawton. Populations, Techniques, Wildlife management. Cervidae.

1172 **Williams, R. W.** 1946. The laboratory rearing of the tropical rat mite, *Liponyssus bacoti* (Hirst). J. Parasitol. 32:252-256.—Tropical rat mite reported to occur on unspecified rodent host in Oklahoma. Parasites. Mammalia.

1173 **Williamson, H. G.** 1956. Habitat improvement evaluation: Segment summary. Okla. Dep. Wildl. Cons. Job 05, W-044-R-05. 4 pp.—Habitat management study includes discussion of *Sylvilagus* and *Odocoileus virginianus.* Natural history, Ecology, Techniques, Wildlife management. Leporidae, Cervidae.

1174 **Williamson, H. G.** 1959. Wildlife habitat improvement: Check survival and adaptability of the planting materials. Okla. Dep. Wildl. Cons. Job 04, W-024-D-14. 5 pp.—Study of *Sigmodon hispidus* populations under various habitat conditions. Natural history, Ecology, Techniques, Economics, Wildlife management. Cricetidae.

1175 **Williamson, H. G., G. B. Wint, and F. F. Copelin.** 1962. Preliminary report on the first four years of the Oklahoma waterfowl and upland game harvest survey, 1958-1961. Proc. Okla. Acad. Sci. 43:242-247.—Game seasons and bag limits. Estimated game kill. Economics, Wildlife management. Leporidae, Sciuridae.

1176 **Willis, L. D., and A. V. Brown.** 1985. Distribution and habitat requirements of the Ozark cavefish, *Amblyopsis rosae.* Am. Midl. Nat. 114:311-317.—Ozark cavefish found in 13 caves in Arkansas, Oklahoma, and Missouri. Noted that gray myotis (*Myotis grisescens*) occupied six of these caves during summer months. Distribution. Vespertilionidae.

1177 **Willner, G. R., G. A. Feldhamer, E. E. Zucker, and J. A. Chapman.** 1980. *Ondatra zibethicus.* Mamm. Species 141:1-8.—*Ondatra zibethi-*

cus cinnamominus in most of Oklahoma, while *O. z. zibethicus* is in eastern and northeastern counties. Summary of information in literature. Systematics, Paleontology, Distribution, Natural history, Behavior, Ecology, Populations, Parasites, Genetics, Morphology, Physiology. Cricetidae.

Willner, G. R., *see* 231.

Wilson, D. E., *see* 356.

Wilson, N., *see* 1157.

1178 **Wilstrand, H.** 1974. Individual, social and seasonal behavior of the thirteen-lined ground squirrel (*Spermophilus tridecemlineatus*). J. Mammal. 55:329-347.—Individual and social behavior evaluated in north-central Texas and south-central Oklahoma. Natural history, Behavior. Sciuridae.

Wimmer, R. B., *see* 1122.

1179 **Winn, B. E., and B. M. Vestal.** 1986. Kin recognition and choice of males by wild female house mice (*Mus musculus*). J. Comp. Psychol. 100:72-75.—Behavior of offspring from wild-trapped animals (parents captured near Norman, Oklahoma) compared with that of laboratory-bred mice. Natural history, Behavior. Muridae.

1180 **Wint, G. B.** 1981. Blue jay killed by fox squirrel. Bull. Okla. Ornithol. Soc. 14:34-35.—*Sciurus niger* killed *Cyanocitta cristata* in Canadian County. Natural history, Behavior. Sciuridae.

Wint, G. B., *see* 628, 629, 1175.

1181 **Wistrand, H.** 1974. Individual, social, and seasonal behavior of the thirteen-lined ground squirrel (*Spermophilus tridecemlineatus*). J. Mammal. 55:329-347.—Study of *Spermophilus tridecemlineatus* in south-central Oklahoma and north-central Texas. Behavior, Ecology. Sciuridae.

1182 **Wittwer, R. F., ed.** 1983. Proceedings—A conference: Creating good wildlife habitat through forest management and conservation. Div. Agric., Okla. State Univ., Stillwater, Oklahoma. 103 pp.—Series of presentations concerning habitats and wildlife. Ecology, Populations, Economics, Wildlife management. Cervidae, Mammalia.

1183 **Wolfe, G., A. A. Kocan, T. R. Thedford, and S. J. Barron.** 1982. Hematologic and serum chemical values of adult female Rocky Mountain elk from New Mexico and Oklahoma. J. Wildl. Dis. 18:223-227.—Hematologic and serum chemical values varied between

Cervus elaphus from New Mexico and those for northeastern Oklahoma. Distribution, Physiology, Wildlife management. Cervidae.

1184 **Wolfe, J. L.** 1982. *Oryzomys palustris.* Mamm. Species 176:1-5.—*Oryzomys palustris texensis* distribution includes eastern Oklahoma. Summary of information in literature. Systematics, Paleontology, Distribution, Natural history, Behavior, Ecology, Parasites, Genetics, Morphology, Physiology, Bibliographies. Cricetidae.

1185 **Wolfe, J. L., and A. V. Linzey.** 1977. *Peromyscus gossypinus.* Mamm. Species 70:1-5.—*Peromyscus gossypinus* distribution includes southeastern Oklahoma. Summary of information in literature. Systematics, Paleontology, Distribution, Natural history, Behavior, Ecology, Populations, Parasites, Genetics, Morphology, Physiology, Bibliographies. Cricetidae.

1186 **Wood, A. E.** 1933. Pleistocene prairie dog from Frederick, Oklahoma. J. Mammal. 14:160.—*Cynomys ludovicianus arizonensis* associated with human artifacts. Paleontology. Sciuridae.

1187 **Woodhouse, S. W.** 1852. The North American jackal—*Canis frustror.* Proc. Acad. Nat. Sci. Phila. 5:147-148.—Type locality of *Canis frustror* (=*latrans*). Systematics. Canidae.

1188 **Woodhouse, S. W.** 1854. Report on the natural history of the country passed over by the exploring expedition under the command of Brev. Capt. L. Sitgreaves, United States topographical engineers, during the year 1851. Mammals. 33rd Congress, 1st Session, Senate Executive Document 59:31-57.—Includes discussion of mammals of Indian Territory. Distribution, Natural history, Behavior, Ecology. Mammalia.

1189 **Woods, C. A.** 1973. *Erethizon dorsatum.* Mamm. Species 29:1-6.—*Erethizon dorsatum* distribution includes northwestern Oklahoma. Summary of information in literature. Systematics, Paleontology, History, Distribution, Natural history, Behavior, Ecology, Populations, Parasites, Genetics, Morphology, Physiology, Economics, Wildlife management, Bibliographies. Erethizontidae.

Wootner, S. J., *see* 968, 969.

1190 **Wyckoff, D. G.** 1980. Caddoan adaptive strategies in the Arkansas Basin, eastern Oklahoma. Ph.D. Dissertation, Washington State Univ., Pullman. 674 pp.—Includes assessment of prehistoric peoples' uses of mammals found in area. Paleontology, Bibliographies. Mammalia.

1191 **Wyckoff, D. G.** 1985. Elephant hunting in Beaver County. Okla. Anthropol. Soc. Newsl. 33(2):7-9.—Prehistoric elephant remains found in Beaver County. Paleontology. Elephantidae.

1192 **Wyman, R. L., and K. Schaefer.** 1972. The ectoparasites of *Peromyscus maniculatus* in Texas County, Oklahoma. Southwest. Nat. 16:435-436.—*Peromyscus maniculatus* ectoparasites include Siphonaptera, Anopleura, Acarina. Five species of fleas. Ecology, Parasites. Cricetidae.

Yates, B. C., *see* 198.

1193 **Yates, T. L., and D. J. Schmidly.** 1975. Karyotype of the eastern mole (*Scalopus aquaticus*), with comments on the karyology of the family Talpidae. J. Mammal. 56:902-905.—Moles, including *Scalopus aquaticus,* exhibit considerable chromosome uniformity. Genetics. Talpidae.

1194 **Yates, T. L., and D. J. Schmidly.** 1977. Systematics of *Scalopus aquaticus* (Linnaeus) in Texas and adjacent states. Occas. Pap. Mus. Tex. Tech Univ. 45:1-36.—*Scalopus aquaticus aereus* occurs in most of Oklahoma, with *S. a. machrinoides* found in northeastern corner. Systematics, Paleontology, Distribution. Talpidae.

1195 **Yates, T. L., and D. J. Schmidly.** 1978. *Scalopus aquaticus.* Mamm. Species 105:1-4.—*Scalopus aquaticus* distribution includes Oklahoma. Summary of information in literature. Systematics, Paleontology, Distribution, Natural history, Behavior, Ecology, Genetics, Morphology, Bibliographies. Talpidae.

Young, C. J., *see* 649.

1196 **Young, S. P.** 1930. Conquering wolfdom and catdom. Southwest Wilds Waters 2:6-7, 47.—Discussion of government predator control measures, primarily against *Canis rufus, C. ater* (=*lupus*), and *C. nubilus* (=*lupus*). History, Distribution, Wildlife management. Canidae, Ursidae, Felidae.

1197 **Young, S. P.** 1946. The wolf in North American history. Caxton Printers, Ltd., Caldwell, Idaho. 149 pp.—*Canis lupus* and *C. niger* (=*rufus*) both occurred in Oklahoma. Includes photograph of red wolf trapped in Carter County. History, Distribution, Economics, Wildlife management, Popular. Canidae.

1198 **Young, S. P.** 1958. The bobcat of North America. Its history, life habits, economic status and control, with list of currently recognized subspecies. Stackpole Co., Harrisburg, Pennsylvania, and Wildlife

Management Institute, Washington, D.C. 193 pp.—*Lynx* (=*Felis*) *rufus* occurs nearly statewide, with *L. r. texensis* in Cimarron County and extreme southwestern Oklahoma. Several localities, with measurements, are listed. Systematics, History, Distribution, Natural history, Behavior, Ecology, Parasites, Morphology, Economics, Wildlife management, Popular. Felidae.

1199 **Young, S. P., and E. A. Goldman.** 1944. The wolves of North America. American Wildlife Institute, Washington, D.C. 636 pp.—*Canis lupus nubilus, Canis niger gregoryi,* and *C. n. rufus* distributions historically included portions of Oklahoma. Systematics, History, Distribution, Natural history, Behavior, Ecology, Parasites, Morphology, Techniques, Economics, Wildlife management, Popular, Bibliographies. Canidae.

1200 **Young, S. P., and E. A. Goldman.** 1946. The puma: Mysterious American cat. American Wildlife Institute, Washington, D.C. 358 pp. —Reports that last known mountain lion (*Felis concolor*) taken in Oklahoma was obtained in 1852. Systematics, History, Distribution, Natural history, Behavior, Ecology, Morphology, Economics, Wildlife management. Felidae.

1201 **Young, S. P., and H. H. T. Jackson.** 1951. The clever coyote. Stackpole Co., Harrisburg, Pennsylvania, and Wildlife Management Institute, Washington, D.C.—Includes "Classification of the races of the coyote," by H. H. T. Jackson. *Canis latrans latrans* in western Oklahoma; *C. l. frustror* in eastern Oklahoma. Systematics, History, Distribution, Natural history, Behavior, Ecology, Parasites, Morphology, Techniques, Economics, Wildlife management. Canidae.

Young, W., *see* 236.

1202 **Zeve, V. H.** 1958. Notes on the biology and distribution of *Trichobius* in northwest Oklahoma (Diptera, Streblidae). Proc. Okla. Acad. Sci. 39:44-49.—Three species of bat-fly *Trichobius* found on *Myotis velifer incautus, Corynorhinus rafinesquei* (=*Plecotus townsendii*) *pallescens,* and *Tadarida brasiliensis mexicana* in Major, Woods, and Woodward counties. Distribution, Parasites. Vespertilionidae, Molossidae.

1203 **Zeve, V. H.** 1959. Additional records of ectoparasitic insects on bats in Oklahoma. Proc. Okla. Acad. Sci. 40:52-56.—Eight species of insect ectoparasites recorded from *Antrozous pallidus, Myotis velifer incautus, Plecotus townsendii pallescens,* and *Tadarida brasiliensis mexicana* from Cimarron, Greer, Harmon, Major, Washita, Woods, and Woodward counties. Distribution, Parasites. Vespertilionidae, Molossidae.

Zim, H. S., *see* 747.

Zimmer, M. A., *see* 607, 609.

1204 **Zimmerman, E. G., B. J. Hart, and C. W. Kilpatrick.** 1975. Biochemical genetics of the *truei* and *boylei* groups of the genus *Peromyscus* (Rodentia). Comp. Biochem. Physiol. B 52:541-545.—Study includes *Peromyscus pectoralis* from Love County. Six species of *Peromyscus* examined. Systematics, Genetics. Cricetidae.

1205 **Zimmerman, E. G., C. W. Kilpatrick, and B. J. Hart.** 1978. The genetics of speciation in the rodent genus *Peromyscus.* Evolution 32:565-579.—Electrophoretic study including specimens of *Peromyscus leucopus* from McCurtain and Cleveland counties, and *P. gossypinus* from McCurtain County. Systematics, Genetics. Cricetidae.

1206 **Zimmerman, E. G., and M. E. Nejtek.** 1977. Genetics and speciation of three semispecies of *Neotoma.* J. Mammal. 58:391-402.—Electrophoretic analysis of *Neotoma albigula, N. micropus,* and *N. floridana.* Systematics, Genetics. Cricetidae.

Zimmerman, E. G., *see* 168, 214, 274, 662, 663, 664, 873.

1207 **Zimmerman, J. W.** 1982. The common long-nosed armadillo (*Dasypus novemcinctus*) in northcentral Oklahoma. M.S. Thesis, Oklahoma State Univ., Stillwater. 106 pp.—Study of *Dasypus novemcinctus* assesses burrow characteristics, home range, foods, and other aspects. Distribution, Natural history, Behavior, Ecology, Morphology, Wildlife management. Dasypodidae.

Zucker, E. E., *see* 1177.

1208 **Zumbaugh, D. M., and J. R. Choate.** 1985. Historical biogeography of foxes in Kansas. Trans. Kans. Acad. Sci. 88:1-13.—Covers distributions of red fox (*Vulpes vulpes*), swift fox (*V. velox*), and gray fox (*Urocyon cinereoargenteus*) in Kansas. Some comments relevant to Oklahoma. History, Distribution, Populations, Ecology. Canidae.

Family and Key Word Index

Family and Key Word Index

Listing of sequence numbers for references in main body of bibliography that pertain to particular mammal family and topic. Families are in alphabetical order, as are the key words under each family. Note that, for a given family, some additional relevant references may be listed under "Mammalia," a general taxonomic designation used when a reference involved more than six families.

Antilocapridae (Pronghorn)

Antilocapridae (Pronghorn) [*continued*]

Bovidae (Bovids)

Bovidae (Bovids) [*continued*]

Camelidae (Camels)

Canidae (Canids)

Canidae (Canids) [*continued*]

Canidae (Canids) [*continued*]

Castoridae (Beavers)

Castoridae (Beavers) [*continued*]

Cervidae (Cervids)

Cervidae (Cervids) [*continued*]

Cervidae (Cervids) [*continued*]

Cricetidae (Cricetids)

Behavior

36 403 509 630 724 821 934 952 1075 1165
56 429 561 631 726 841 935 953 1132 1166
69 433 580 632 729 853 937 1015 1133 1177
91 440 588 666 766 897 938 1017 1139 1184
216 451 592 695 769 909 947 1018 1151 1185
271 452 594 696 820 911

Bibliographies

216 588 695 766 953 1151 1165 1166 1184 1185
271 666 724 952 1018

Distribution

6 214 359 441 571 605 724 853 952 1073
69 216 360 443 580 612 759 856 953 1075
90 232 387 475 587 621 760 864 1005 1086
91 233 400 477 588 632 761 887 1010 1122
103 242 403 481 589 641 766 890 1015 1153
141 257 417 509 594 648 803 897 1017 1165
144 265 428 523 595 661 809 910 1018 1166
146 271 433 552 596 666 820 934 1021 1177
147 312 436 558 603 695 821 941 1022 1184
161 340 438 566 604 696 833 950 1024 1185
196 341 440

Ecology

36 216 403 449 592 661 768 897 953 1132
56 232 417 455 594 666 769 909 1015 1151
69 269 428 481 604 695 780 911 1017 1162
91 271 429 509 612 696 809 924 1018 1165
117 301 431 552 621 724 820 934 1028 1166
142 312 433 556 628 726 821 935 1029 1174
144 342 435 580 630 729 841 946 1032 1177
153 360 438 585 631 760 842 947 1075 1184
160 373 439 586 632 761 853 950 1106 1185
215 387 440 588 641 766 878 952 1109 1192

Economics

91 403 475 594 628 769 841 853 1028 1032
216 440 543 612 684 820 842 947 1029 1174
301 441

Cricetidae (Cricetids) [*continued*]

Cricetidae (Cricetids) [*continued*]

Dasypodidae (Armadillos)

Dasypodidae (Armadillos) [*continued*]

Dasypodidae (Armadillos) [*continued*]

Didelphidae (Opossums)

Didelphidae (Opossums) [*continued*]

Dromomerycidae (Dromomerycids)

Elephantidae (Elephants and Mammoths)

Elephantidae (Elephants and Mammoths) [*continued*]

Equidae (Horses)

Erethizontidae (New World Porcupines)

Erethizontidae (New World Porcupines) [*continued*]

Felidae (Cats)

Felidae (Cats) [*continued*]

Distribution
58 181 348 523 544 709 719 943 1011 1198
119 286 486 524 618 710 783 960 1060 1200
145 287 520 535 671 711 797 978 1196

Ecology
58 287 383 628 719 780 924 1033 1060 1198
181 339 530 629 727 782 978 1037 1162 1200
286 380 627 710 730 797

Economics
286 522 627 629 804 923 1033 1060 1198 1200
339 543 628 719

Genetics
286 1161

History
286 486 524 543 719 1011 1060 1196 1198 1200

Morphology
58 286 710 956 978 1033 1092 1161 1198 1200

Natural History
58 279 383 627 719 783 957 1033 1060 1198
112 286 530 628 727 924 978 1037 1162 1200
181 380 618 629

Paleontology
246 286 670 671 672 749 790 918 943 963
268

Parasites
286 427 459 680 719 782 974 1033 1162 1198
383

Physiology
776

Popular
58 181 348 486 618 710 978 1011 1060 1198
119

Felidae (Cats) [*continued*]

Geomyidae (Pocket Gophers)

Geomyidae (Pocket Gophers) [*continued*]

Glyptodontidae (Glyptodonts)

Gomphotheriidae (Gomphotheriids)

Heteromyidae (Heteromyids)

Heteromyidae (Heteromyids) [*continued*]

Leporidae (Hares and Rabbits)

Leporidae (Hares and Rabbits) [*continued*]

Leporidae (Hares and Rabbits) [*continued*]

Mammalia (Multifamily Citations)

Mammalia (Multifamily Citations) [*continued*]

Mammalia (Multifamily Citations) [*continued*]

Mastodontidae (Mastodons)

Mastodontidae (Mastodons) [continued]

Megalonychidae (Ground Sloths)

Molossidae (Molossid Bats)

Molossidae (Molossid Bats) [*continued*]

Muridae (Murids)

Muridae (Murids) [*continued*]

Mustelidae (Mustelids)

Mustelidae (Mustelids) [*continued*]

Mustelidae (Mustelids) [*continued*]

Mylagaulidae (Mylagaulids)

Mylodontidae (Mylodonts)

Myocastoridae (Myocastorids)

Myocastoridae (Myocastorids) [*continued*]

Procyonidae (Procyonids)

Procyonidae (Procyonids) [*continued*]

Rhinocerotidae (Rhinoceroses)

Rhinocerotidae (Rhinoceroses) [*continued*]

Sciuridae (Squirrels)

Sciuridae (Squirrels) [*continued*]

Sciuridae (Squirrels) [*continued*]

Soricidae (Shrews)

Soricidae (Shrews) [*continued*]

Talpidae (Moles)

Talpidae (Moles) [*continued*]

Tapiridae (Tapirs)

Tayassuidae (Peccaries)

Ursidae (Bears)

Ursidae (Bears) [*continued*]

Vespertilionidae (Vespertilionid Bats)

Vespertilionidae (Vespertilionid Bats) [*continued*]

Distribution

3 191 362 423 554 691 807 857 1086 1127
21 206 363 426 564 692 808 884 1087 1141
33 209 376 456 622 693 812 896 1096 1145
98 298 389 476 623 700 813 1000 1098 1147
104 324 407 507 636 701 817 1001 1099 1176
145 353 408 510 658 748 825 1009 1111 1202
152 357 412 517 687 798 847 1064 1112 1203
188 358 415 518 690 801

Ecology

104 209 362 408 456 622 691 817 1064 1099
152 213 363 415 507 623 692 825 1087 1145
187 324 376 416 517 636 693 1000 1096 1147
206 353 389 423 564 690 808 1001 1098

Economics

104 376 564 817

Genetics

353 376 564 636 692 693 1000 1001 1087 1147
363

History

21 376 510 564 623 690 798 884 1112

Morphology

3 187 362 424 507 690 748 884 1087 1141
98 188 363 425 510 692 798 1000 1096 1145
104 353 376 426 564 693 812 1001 1127 1147
182 358 389 476 636 701 817

Natural History

3 209 357 407 517 623 691 825 1001 1098
104 324 362 415 554 636 692 884 1087 1099
152 352 363 416 564 653 693 915 1096 1111
206 353 376 456 619 690 817 1000 1097 1147

Paleontology

104 363 376 507 564 636 692 693 1000 1001
353

Vespertilionidae (Vespertilionid Bats) [*continued*]

Zapodidae (Jumping Mice)

Zapodidae (Jumping Mice) [*continued*]

English Name Index

English Name Index

Alphabetical listing of English common names of Oklahoma mammals. Page number is indicated for appropriate section of Family and Key Word Index that lists Oklahoma references covering a particular type or group of mammals. Where family name is not included below, the common name as listed here also appears on page indicated in Family and Key Word Index.